cord.

THE STRATEGY OF CONFLICT

THE STRATEGY

OF CONFLICT

THOMAS C. SCHELLING

PROFESSOR OF ECONOMICS AND
ASSOCIATE, CENTER FOR INTERNATIONAL AFFAIRS
HARVARD UNIVERSITY

HARVARD UNIVERSITY PRESS
CAMBRIDGE, MASSACHUSETTS

1960

PREFACE

This is a series of closely interrelated essays in a field that is variously described as "theory of bargaining," "theory of conflict," or "theory of strategy." Strictly speaking, the subject falls within the *theory of games*, but within the part of game theory in which the least satisfactory progress has been made, the situations in which there is common interest as well as conflict between adversaries: negotiations, war and threats of war, criminal deterrence, tacit bargaining, extortion. The philosophy of the book is that in the strategy of conflict there are enlightening similarities between, say, maneuvering in limited war and jockeying in a traffic jam, between deterring the Russians and deterring one's own children, or between the modern balance of terror and the ancient institution of hostages.

The analysis is neither difficult nor so dependent on mathematics or analytical apparatus as to be inaccessible to any serious reader. A few chapters call for a rudimentary acquaintance with some concepts from game theory.

The first chapter (in a longer version) was originally presented in early 1959 to a conference on "International Relations in the Mid-twentieth Century," at Northwestern University; although the occasion and the audience were somewhat specialized, the paper represents the motivation and theme of the entire book. Chapters 2 and 3 were originally independent articles on "bargaining." It was evident, after they were written, that they belonged to the same field as the *theory of games*; an effort to fit them into the framework of game theory, stretching the framework if necessary, resulted in Chapters 4 through 6 and Appendices B and C. Chapters 7 through 10, and Appendix A, are extensions of the same method to particular problems in international strategy.

Appendices B and C will be of interest mainly to readers conversant with bargaining theory or game theory. Appendix A has been treated as an appendix only because its extended preoccupation with a particular policy problem is in some contrast to the style of Chapter 4, where it would otherwise belong.

The essays are a mixture of "pure" and "applied" research. To some extent the two can be separated, as in the companion pieces in Part IV. In my own thinking they have never been separate. Motivation for the purer theory came almost exclusively from preoccupation with (and fascination with) "applied" problems; and the clarification of theoretical ideas was absolutely dependent on an identification of live examples. For reasons inherent either in the subject or in the author, the interaction of the two levels of theory has been continuous and intense.

Three people have been most influential, probably more than they realize, in my continuing this work. They are Kenneth E. Boulding, Bernard F. Haley, and Charles J. Hitch. Numerous associates, particularly at The RAND Corporation, have lent me ideas and stimulated my own; I refer especially to Bernard Brodie, Daniel Ellsberg, Malcolm W. Hoag, Herman Kahn, William W. Kaufmann, and Albert J. Wohlstetter. William W. Taylor gave me valuable editorial help. And I owe a special word of appreciation to R. Duncan Luce and Howard Raiffa, whose *Games and Decisions* has been of immeasurable help; if I have often focused critical remarks on the book, it is only because the inevitable lot of a definitive survey is to serve as a definitive target.

During the year before this book went to press I was uniquely located to receive stimulation, provocation, advice, comment, disagreement, encouragement, and education. I spent the year with The RAND Corporation, in Santa Monica. As a collection of people, RAND is superb, and I have mentioned above only the few whose intellectual impact on me was powerful and persistent; many others, truly too numerous to list here, have as individuals affected the final shape of this book. But RAND is more than a collection of people; it is a social organism characterized by intellect, imagination, and good humor. RAND is not responsible for the shapes my ideas have taken — the "views herein expressed" — but I hope it will, as a corporation, take satisfaction from its responsibility for some of the ideas' taking any shape at all.

For readers who have come across some of the chapters before, the following may be of convenience. Chapter 2 appeared with the same title in *The American Economic Review*, Vol. XLVI No. 3, June 1956. Chapter 3 appeared with the same title in *The Journal of Conflict Resolution*, Vol. I No. 1, March 1957. Chapters 4, 5, and 6 are a somewhat rearranged version of "The Strategy of Conflict," *The Journal of Conflict Resolution*, Vol. II No. 3, September 1958, with parts eliminated that overlapped other chapters. Appendix B appeared, with the same title, in *The Review of Economics and Statistics*, Vol. XLI No. 3, August 1959. A longer version of Chapter 10, with the same title, is contained in Klaus Knorr (ed.), *NATO and American Security*, (Princeton: Princeton University Press, 1959). The several publishers have kindly allowed me to reprint these papers here, with modifications to make an integrated book.

THOMAS C. SCHELLING

Cambridge, Massachusetts

CONTENTS

PART I

ELEMENTS OF A
THEORY OF STRATEGY

1

THE RETARDED SCIENCE OF
INTERNATIONAL STRATEGY

Among diverse theories of conflict — corresponding to the diverse meanings of the word "conflict" — a main dividing line is between those that treat conflict as a pathological state and seek its causes and treatment, and those that take conflict for granted and study the behavior associated with it. Among the latter there is a further division between those that examine the participants in a conflict in all their complexity — with regard to both "rational" and "irrational" behavior, conscious and unconscious, and to motivations as well as to calculations — and those that focus on the more rational, conscious, artful kind of behavior. Crudely speaking, the latter treat conflict as a kind of contest, in which the participants are trying to "win." A study of conscious, intelligent, sophisticated conflict behavior — of successful behavior — is like a search for rules of "correct" behavior in a contest-winning sense.

We can call this field of study the *strategy* of conflict.[1] We can be interested in it for at least three reasons. We may be involved in a conflict ourselves; we all are, in fact, participants in international conflict, and we want to "win" in some proper sense. We may wish to understand how participants actually do conduct themselves in conflict situations; an understanding of "correct" play may give us a bench mark for the study of actual behavior.

[1] The term "strategy" is taken, here, from the *theory of games*, which distinguishes games of skill, games of chance, and games of strategy, the latter being those in which the best course of action for each player depends on what the other players do. The term is intended to focus on the interdependence of the adversaries' decisions and on their expectations about each other's behavior. This is not the military usage.

We may wish to control or influence the behavior of others in conflict, and we want, therefore, to know how the variables that are subject to our control can affect their behavior.

If we confine our study to the theory of strategy, we seriously restrict ourselves by the assumption of rational behavior — not just of intelligent behavior, but of behavior motivated by a conscious calculation of advantages, a calculation that in turn is based on an explicit and internally consistent value system. We thus limit the applicability of any results we reach. If our interest is the study of actual behavior, the results we reach under this constraint may prove to be either a good approximation of reality or a caricature. Any abstraction runs a risk of this sort, and we have to be prepared to use judgment with any results we reach.

The advantage of cultivating the area of "strategy" for theoretical development is not that, of all possible approaches, it is the one that evidently stays closest to the truth, but that the assumption of rational behavior is a productive one. It gives a grip on the subject that is peculiarly conducive to the development of theory. It permits us to identify our own analytical processes with those of the hypothetical participants in a conflict; and by demanding certain kinds of consistency in the behavior of our hypothetical participants, we can examine alternative courses of behavior according to whether or not they meet those standards of consistency. The premise of "rational behavior" is a potent one for the production of theory. Whether the resulting theory provides good or poor insight into actual behavior is, I repeat, a matter for subsequent judgment.

But, in taking conflict for granted, and working with an image of participants who try to "win," a theory of strategy does not deny that there are common as well as conflicting interests among the participants. In fact, the richness of the subject arises from the fact that, in international affairs, there is mutual dependence as well as opposition. Pure conflict, in which the interests of two antagonists are completely opposed, is a special case; it would arise in a war of complete extermination, otherwise not even in war. For this reason, "winning" in a conflict does not have a strictly competitive meaning; it is not winning relative to one's adversary. It means gaining relative to one's own value system;

and this may be done by bargaining, by mutual accommodation, and by the avoidance of mutually damaging behavior. If war to the finish has become inevitable, there is nothing left but pure conflict; but if there is any possibility of avoiding a mutually damaging war, of conducting warfare in a way that minimizes damage, or of coercing an adversary by threatening war rather than waging it, the possibility of mutual accommodation is as important and dramatic as the element of conflict. Concepts like deterrence, limited war, and disarmament, as well as negotiation, are concerned with the common interest and mutual dependence that can exist between participants in a conflict.

Thus, strategy — in the sense in which I am using it here — is not concerned with the efficient *application* of force but with the *exploitation of potential force*. It is concerned not just with enemies who dislike each other but with partners who distrust or disagree with each other. It is concerned not just with the division of gains and losses between two claimants but with the possibility that particular outcomes are worse (better) for *both* claimants than certain other outcomes. In the terminology of game theory, most interesting international conflicts are not "constant-sum games" but "variable-sum games": the sum of the gains of the participants involved is not fixed so that more for one inexorably means less for the other. There is a common interest in reaching outcomes that are mutually advantageous.

To study the strategy of conflict is to take the view that most conflict situations are essentially *bargaining* situations. They are situations in which the ability of one participant to gain his ends is dependent to an important degree on the choices or decisions that the other participant will make. The bargaining may be explicit, as when one offers a concession; or it may be by tacit maneuver, as when one occupies or evacuates strategic territory. It may, as in the ordinary haggling of the market-place, take the *status quo* as its zero point and seek arrangements that yield positive gains to both sides; or it may involve threats of damage, including mutual damage, as in a strike, boycott, or price war, or in extortion.

Viewing conflict behavior as a bargaining process is useful in keeping us from becoming exclusively preoccupied either with the

conflict or with the common interest. To characterize the maneuvers and actions of limited war as a bargaining process is to
emphasize that, in addition to the divergence of interest over the
variables in dispute, there is a powerful common interest in reaching an outcome that is not enormously destructive of values to
both sides. A "successful" employees' strike is not one that
destroys the employer financially, it may even be one that never
takes place. Something similar can be true of war.

The idea of "deterrence" has had an evolution that is instructive for our purpose. It is a dozen years since deterrence was ar-
ticulated as the keystone of our national strategy, and during
those years the concept has been refined and improved. We have
learned that a threat has to be credible to be efficacious, and that
its credibility may depend on the costs and risks associated with
fulfillment for the party making the threat. We have developed
the idea of making a threat credible by getting ourselves committed to its fulfillment, through the stretching of a "trip wire"
across the enemy's path of advance, or by making fulfillment a
matter of national honor and prestige — as in the case, say, of
the Formosa Resolution. We have recognized that a readiness to
fight limited war in particular areas may detract from the threat
of massive retaliation, by preserving the choice of a lesser evil if
the contingency arises. We have considered the possibility that a
retaliatory threat may be more credible if the means of carrying
it out and the responsibility for retaliation are placed in the hands
of those whose resolution is strongest, as in recent suggestions for
"nuclear sharing." We have observed that the rationality of the
adversary is pertinent to the efficacy of a threat, and that mad-
men, like small children, can often not be controlled by threats.
We have recognized that the efficacy of the threat may depend
on what alternatives are available to the potential enemy, who,
if he is not to react like a trapped lion, must be left some tolerable
recourse. We have come to realize that a threat of all-out retaliation gives the enemy every incentive, in the event he should
choose not to heed the threat, to initiate his transgression with an
all-out strike at us; it eliminates lesser courses of action and
forces him to choose between extremes. We have learned that the

threat of massive destruction may deter an enemy only if there is a corresponding implicit promise of nondestruction in the event he complies, so that we must consider whether too great a capacity to strike him by surprise may induce him to strike first to avoid being disarmed by a first strike from us. And recently, in connection with the so-called "measures to safeguard against surprise attack," we have begun to consider the possibility of improving mutual deterrence through arms control.

What is impressive is not how complicated the idea of deterrence has become, and how carefully it has been refined and developed, but how slow the process has been, how vague the concepts still are, and how inelegant the current theory of deterrence is. This is not said to depreciate the efforts of people who have struggled with the deterrence concept over the last dozen years. On strategic matters of which deterrence is an example, those who have tried to devise policies to meet urgent problems have had little or no help from an already existing body of theory, but have had to create their own as they went along. There is no scientific literature on deterrence that begins to compare with, say, the literature on inflation, Asiatic flu, elementary-school reading, or smog.

Furthermore, those who have grappled with ideas like deterrence, being motivated largely by immediate problems, have not primarily been concerned with the cumulative process of developing a theoretical structure. This seems to be true not only of policy-makers and journalists but of the more scholarly as well. Whether it reflects the scholars' interests or that of the editors, the literature on deterrence and related concepts has been mainly preoccupied with solving immediate problems rather than with a methodology for dealing with problems.[2] We do not even have a

---

[2] There are some excellent examples to the contrary, like C. W. Sherwin, "Securing Peace Through Military Technology," *Bulletin of the Atomic Scientists*, 12:159–164 (May 1956). And Sherwin's reference there to a paper by Warren Amster reminds us that when theory is stimulated by military problems, as so much of it currently is, it may not receive open publication. There are undoubtedly, also, serious editorial obstacles; journals in international affairs appeal to a dominantly nontheoretical audience, and articles with high theoretical content must often be purged of it and focused on immediate problems. The recent devotion of an entire issue of *Conflict Resolution* to Anatol Rapoport's magnificent essay on "Lewis F. Richardson's Mathematical

decent terminology; occasional terms like "active" and "passive" deterrence do not begin to fill the need.

How do we account for this lack of theoretical development? I think one significant fact is that the military services, in contrast to almost any other sizable and respectable profession, have no identifiable academic counterpart. Those who make policy in the fields of economics, medicine, public health, soil conservation, education, or criminal law, can readily identify their scholarly counterpart in the academic world. (In economics the number of trained people who are doing research and writing books compares well with the number engaged in economic policy or administration.) But where is the academic counterpart of the military profession?

It is not — on any great scale — in the service academies; these are undergraduate schools, devoted mainly to teaching rather than to research. Not — or not yet on any great scale — in the war colleges and other nontechnical advanced educational institutions within the military services; these have not yet developed the permanent faculty, the research orientation, and the value system required for sustained and systematic theoretical development.

Within the universities, military strategy in this country has been the preoccupation of a small number of historians and political scientists, supported on a scale that suggests that deterring the Russians from a conquest of Europe is about as important as enforcing the antitrust laws. This is said not to disparage the accomplishments, but to emphasize that within the universities there has usually been no directly identifiable department or line of inquiry that can be associated with the military professions and the role of force in foreign relations. (ROTC programs have recently become a limited exception to this point, at least to the extent that they induce the organization of pertinent courses in history and political science.) The defense-studies programs and institutes now found on a number of campuses, and the attention given to international security problems by the foundations, are a novel and significant development. New quasi-governmental

---

Theory of War" (vol. I, No. 3, September 1957) is a heartening sign in the other direction.

research institutions like The RAND Corporation and the Institute for Defense Analysis are importantly helping to fill the need but, for our purpose, can be cited as evidence of the need.

One may ask whether the military services themselves might not be able to produce a growing body of theory to illuminate ideas like deterrence or limited war. After all, theory does not have to be developed solely by specialists isolated in universities. If the military services are intellectually prepared to make effective *use* of military force, it might seem that they are equipped to theorize about it. But here a useful distinction can be made between the *application* of force and the *threat* of force. Deterrence is concerned with the exploitation of potential force. It is concerned with persuading a potential enemy that he should in his own interest avoid certain courses of activity. There is an important difference between the intellectual skills required for carrying out a military mission and for using *potential* military capability to pursue a nation's objectives. A theory of deterrence would be, in effect, a theory of the skillful *nonuse* of military forces, and for this purpose deterrence requires something broader than military skills. The military professions may have these broader skills, but they do not automatically have them as a result of meeting their primary responsibilities, and those primary responsibilities place full-time demands on their time.[3]

A new kind of inquiry that gave promise, fifteen years ago, of leading to such a theory of strategy is *game theory*. Game theory is concerned with situations — games of "strategy," in contrast to games of skill or games of chance — in which the best course of action for each participant depends on what he expects the

---

[3] The lack of a vigorous intellectual tradition in the field of military strategy is forcefully discussed by Bernard Brodie in the first chapters of his *Strategy in the Missile Age* (Princeton, 1959). Pertinent also is Colonel Joseph I. Greene's foreword to the Modern Library edition of Clausewitz, *On War* (New York, 1943): "During most of the years between the great wars, the two highest schools of our Army were limited to a single course of some ten months' duration for all officers selected to attend them. . . . There could be no time at either place for study of the long development of military thought and theory. . . . If ever more extensive periods of higher training become possible in our Army — periods of two or three years' duration — the greatest of the military thinkers would surely deserve a course of study in themselves" (pp. xi–xii).

other participants to do. A deterrent threat meets this definition nicely; it works only because of what the other player expects us to do in response to his choice of moves, and we can afford to make the threat only because we expect it to have an influence on his choice. But in international strategy the promise of game theory is so far unfulfilled. Game theory has been extremely helpful in the formulation of problems and the clarification of concepts, but its greatest successes have been in other fields. It has, on the whole, been pitched at a level of abstraction where it has made little contact with the elements of a problem like deterrence.[4]

The idea of deterrence figures so prominently in some areas of conflict other than international affairs that one might have supposed the existence of a well-cultivated theory already available to be exploited for international applications. Deterrence has been an important concept in criminal law for a long time. Legislators, jurists, lawyers, and legal scholars might be supposed to have subjected the concept to rigorous and systematic scrutiny for many generations. To be sure, deterrence is not the sole consideration involved in criminal law, nor even necessarily the most important; still, it has figured prominently enough for one to suppose the existence of a theory that would take into account the kinds and sizes of penalties available to be imposed on a convicted criminal, the potential criminal's value system, the profitability of crime, the law-enforcement system's ability to apprehend criminals and to get them convicted, the criminal's awareness of the law and of the probability of apprehension and conviction, the extent to which different types of crime are motivated by rational calculation, the resoluteness of society to be neither niggardly nor soft-hearted in the expensive and disagreeable application of the penalty and how well this reso-

---

[4] Jessie Bernard, writing on "The Theory of Games as a Modern Sociology of Conflict," gives a somewhat similar appraisal but adds that "we may expect that the mathematics required to make a fruitful application of the theory of games to sociological phenomena will emerge in the not-too-distant future" (*The American Journal of Sociology*, 59:418, March 1954). My own view is that the present deficiencies are not in the mathematics, and that the theory of strategy has suffered from too great a willingness of social scientists to treat the subject as though it were, or should be, solely a branch of mathematics.

luteness (or lack of it) is known to the criminal, the likelihood of mistakes in the system, the possibilities for third parties to exploit the system for personal gain, the role of communication between organized society and the criminal, the organization of criminals to defeat the system, and so on.

It is not only criminals, however, but our own children that have to be deterred. Some aspects of deterrence stand out vividly in child discipline: the importance of rationality and self-discipline on the part of the person to be deterred, of his ability to comprehend the threat if he hears it and to hear it through the din and noise, of the threatener's determination to fulfill the threat if need be — and, more important, of the threatened party's conviction that the threat will be carried out. Clearer perhaps in child discipline than in criminal deterrence is the important possibility that the threatened punishment will hurt the threatener as much as it will the one threatened, perhaps more. There is an analogy between a parent's threat to a child and the threat that a wealthy paternalistic nation makes to the weak and disorganized government of a poor nation in, say, extending foreign aid and demanding "sound" economic policies or cooperative military policies in return.

And the analogy reminds us that, even in international affairs, deterrence is as relevant to relations between friends as between potential enemies. (The threat to withdraw to a "peripheral strategy" if France failed to ratify the European Defense Community Treaty was subject to many of the same disabilities as a threat of retaliation.) The deterrence concept requires that there be both conflict and common interest between the parties involved; it is as inapplicable to a situation of pure and complete antagonism of interest as it is to the case of pure and complete common interest. Between these extremes, deterring an ally and deterring an enemy differ only by degrees, and in fact we may have to develop a more coherent theory before we can even say in a meaningful way whether we have more in common with Russia or with Greece, relative to the conflicts between us.[5]

---

[5] It may be important to emphasize that, in referring to a "common interest," I do not mean that they must have what is usually referred to as a similarity in their value systems. They may just be in the same boat together.

The deterrence idea also crops up casually in everyday affairs. Automobile drivers have an evident common interest in avoiding collision and a conflict of interest over who shall go first and who shall slam on his brakes and let the other through. Collision being about as mutual as anything can be, and often the only thing that one can threaten, the maneuvers by which one conveys a threat of mutual damage to another driver aggressing on one's right of way are an instructive example of the kind of threat that is conveyed not by words but by actions, and of the threat in which the pledge to fulfill is made not by verbal announcement but by losing the power to do otherwise.

Finally, there is the important area of the underworld. Gang war and international war have a lot in common. Nations and outlaws both lack enforceable legal systems to help them govern their affairs. Both engage in the ultimate in violence. Both have an interest in avoiding violence, but the threat of violence is continually on call. It is interesting that racketeers, as well as gangs of delinquents, engage in limited war, disarmament and disengagement, surprise attack, retaliation and threat of retaliation; they worry about "appeasement" and loss of face; and they make alliances and agreements with the same disability that nations are subject to — the inability to appeal to higher authority in the interest of contract enforcement.

There are consequently a number of other areas available for study that may yield insight into the one that concerns us, the international area. Often a principle that in our own field of interest is hidden in a mass of detail, or has too complicated a structure, or that we cannot see because of a predisposition, is easier to perceive in another field where it enjoys simplicity and vividness or where we are not blinded by our predispositions. It may be easier to articulate the peculiar difficulty of constraining

---

They may even be there only because one of them perceived it a strategic advantage to get in that position — to couple their interests in not tipping the boat. If being overturned together in the same boat is a potential outcome, given the array of alternatives available to both parties, they have a "common interest" in the sense intended in the text. "Potential common interest" might seem more descriptive. Deterrence, for example, is concerned with coupling one's own course of action with the other's course of action in a way that exploits that potential common interest.

a Mossadeq by the use of threats when one is fresh from a vain attempt at using threats to keep a small child from hurting a dog or a small dog from hurting a child.

None of these other areas of conflict seems to have been mastered by a well-developed theory than can, with modification, be used in the analysis of international affairs. Sociologists, including those who study criminal behavior in underworld conflict, have not traditionally been much concerned with what we would call the *strategy* of conflict. Nor does the literature on law and criminology reveal an appreciable body of explicit theory on the subject. I cannot confidently assert that there are no handbooks, textbooks, or original works on the pure theory of blackmail circulating in the underworld; but certainly no expurgated version, showing how to use extortion and how to resist it, has shown up as "New Ways in Child Guidance," in spite of the demand for it.[6]

What would "theory" in this field of strategy consist of? What questions would it try to answer? What ideas would it try to unify, clarify, or communicate more effectively? To begin with, it should define the essentials of the situation and of the behavior in question. Deterrence — to continue with deterrence as a typical strategic concept — is concerned with influencing the choices that another party will make, and doing it by influencing his expectations of how we will behave. It involves confronting him with evidence for believing that our behavior will be determined by his behavior.

But what configuration of value systems for the two participants — of the "payoffs," in the language of game theory — makes a deterrent threat credible? How do we measure the mixture of conflict and common interest required to generate a "deterrence" situation? What communication is required, and what means of authenticating the evidence communicated? What kind of "rationality" is required of the party to be deterred — a knowledge of his own value system, an ability to perceive alternatives

[6] Progress is being made. Daniel Ellsberg included a lecture on "The Theory and Practice of Blackmail," and one on "The Political Uses of Madness," in his series on "The Art of Coercion," sponsored by the Lowell Institute, Boston, March 1959.

and to calculate with probabilities, an ability to demonstrate (or an inability to conceal) his own rationality?

What is the need for trust, or enforcement of promises? Specifically, in addition to threatening damage, need one also guarantee to withhold the damage if compliance is forthcoming; or does this depend on the configuration of "payoffs" involved? What "legal system," communication system, or information structure is needed to make the necessary promises enforceable?

Can one threaten that he will "probably" fulfill a threat; or must he threaten that he certainly will? What is the meaning of a threat that one will "probably" fulfill when it is clear that, if he retained any choice, he'd have no incentive to fulfill it after the act? More generally, what are the devices by which one gets committed to fulfillment that he would otherwise be known to shrink from, considering that if a commitment makes the threat credible enough to be effective it need not be carried out. What is the difference, if any, between a threat that deters action and one that compels action, or a threat designed to safeguard a second party from his own mistakes? Are there any logical differences among deterrent, disciplinary, and extortionate threats?

How is the situation affected by a third participant, who has his own mixture of conflict and common interest with those already present, who has access to or control of the communication system, whose behavior is rational or irrational in one sense or another, who enjoys trust or some means of contract enforcement with one or another of the two principals? How are these questions affected by the existence of a legal system that permits and prohibits certain actions, that is available to inflict penalty on nonfulfillment of contract, or that can demand authentic information from the participants. To what extent can we rationalize concepts like "reputation," "face," or "trust," in terms of a real or hypothetical legal system, in terms of modification of the participants' value systems, or in terms of relationships of the players concerned to additional participants, real or hypothetical?

This brief sample of questions may suggest that there is scope for the creation of "theory." There is something here that looks like a mixture of game theory, organization theory, communication theory, theory of evidence, theory of choice, and theory of collective decision. It is faithful to our definition of "strategy": it takes conflict for granted, but also assumes common interest between the adversaries; it assumes a "rational" value-maximizing mode of behavior; and it focuses on the fact that each participant's "best" choice of action depends on what he expects the other to do, and that "strategic behavior" is concerned with influencing another's choice by working on his expectation of how one's own behavior is related to his.

There are two points worth stressing. One is that, though "strategy of conflict" sounds cold-blooded, the theory is not concerned with the efficient *application* of violence or anything of the sort; it is not essentially a theory of aggression or of resistance or of war. *Threats* of war, yes, or threats of anything else; but it is the employment of threats, or of threats and promises, or more generally of the conditioning of one's own behavior on the behavior of others, that the theory is about.

Second, such a theory is nondiscriminatory as between the conflict and the common interest, as between its applicability to potential enemies and its applicability to potential friends. The theory degenerates at one extreme if there is no scope for mutual accommodation, no common interest at all even in avoiding mutual disaster; it degenerates at the other extreme if there is no conflict at all and no problem in identifying and reaching common goals. But in the area between those two extremes the theory is noncommittal about the mixture of conflict and common interest; we can equally well call it the theory of precarious partnership or the theory of incomplete antagonism.[7] (In Chapter 9 it is pointed out that some central aspects of the problem of surprise attack in international affairs are structurally identical with the problem of mutually suspicious partners.)

Both of these points — the neutrality of the theory with respect to the degree of conflict involved, and the definition of "strategy" as concerned with constraining an adversary through

[7] In using the word "threat" I have not intended any necessarily aggressive or hostile connotations. In an explicit negotiation between friends or in tacit cooperation between them, the threat of disagreement or of reduced cooperation, expressed or implied, is a sanction by which they support their demands, just as in a commercial transaction an offer is enforced by threat of "no sale."

his expectation of the consequences of his actions — suggest that we might call our subject the *theory of interdependent decision.*

Threats and responses to threats, reprisals and counter-reprisals, limited war, arms races, brinkmanship, surprise attack, trusting and cheating can be viewed as either hot-headed or cool-headed activities. In suggesting that they can usefully be viewed, in the development of theory, as cool-headed activities, it is not asserted that they are in fact entirely cool-headed. Rather it is asserted that the assumption of rational behavior is a productive one in the generation of systematic theory. If behavior were actually cool-headed, valid and relevant theory would probably be easier to create than it actually is. If we view our results as a bench mark for further approximation to reality, not as a fully adequate theory, we should manage to protect ourselves from the worst results of a biased theory.

Furthermore, theory that is based on the assumption that the participants coolly and "rationally" calculate their advantages according to a consistent value system forces us to think more thoroughly about the meaning of "irrationality." Decision-makers are not simply distributed along a one-dimensional scale that stretches from complete rationality at one end to complete irrationality at the other. Rationality is a collection of attributes, and departures from complete rationality may be in many different directions. Irrationality can imply a disorderly and inconsistent value system, faulty calculation, an inability to receive messages or to communicate efficiently; it can imply random or haphazard influences in the reaching of decisions or the transmission of them, or in the receipt or conveyance of information; and it sometimes merely reflects the collective nature of a decision among individuals who do not have identical value systems and whose organizational arrangements and communication systems do not cause them to act like a single entity.

As a matter of fact, many of the critical elements that go into a model of rational behavior can be identified with particular types of rationality or irrationality. The value system, the communication system, the information system, the collective decision process, or a parameter representing the probability of error or loss of control, can be viewed as an effort to formalize the study of "irrationality." Hitler, the French Parliament, the commander of a bomber, the radar operators at Pearl Harbor, Khrushchev, and the American electorate may all suffer from some kinds of "irrationality," but by no means the same kinds. Some of them can be accounted for within a theory of rational behavior. (Even the neurotic, with inconsistent values and no method of reconciling them, motivated to suppress rather than to reconcile his conflicting goals, may for some purposes be viewed as a *pair* of "rational" entities with distinct value systems, reaching collective decisions through a voting process that has some haphazard or random element, asymmetrical communications, and so forth.)

The apparent restrictiveness of an assumption of "rational" behavior — of a calculating, value-maximizing strategy of decision — is mitigated by two additional observations. One, which I can only allege at second hand, is that even among the emotionally unbalanced, among the certified "irrationals," there is often observed an intuitive appreciation of the principles of strategy, or at least of particular applications of them. I am told that inmates of mental hospitals often seem to cultivate, deliberately or instinctively, value systems that make them less susceptible to disciplinary threats and more capable of exercising coercion themselves. A careless or even self-destructive attitude toward injury — "I'll cut a vein in my arm if you don't let me . . ." — can be a genuine strategic advantage; so can a cultivated inability to hear or to comprehend, or a reputation for frequent lapses of self-control that make punitive threats ineffectual as deterrents. (Again I am reminded of my children.) As a matter of fact, one of the advantages of an explicit theory of "rational" strategic decision in situations of mixed conflict and common interest is that, by showing the strategic basis of certain paradoxical tactics, it can display how sound and rational some of the tactics are that are practiced by the untutored and the infirm. It may not be an exaggeration to say that our sophistication sometimes suppresses sound intuitions, and one of the effects of an explicit theory may be to restore some intuitive notions that were only superficially "irrational."

The second observation is related to the first. It is that an explicit theory of "rational" decision, and of the strategic consequences of such decisions, makes perfectly clear that it is not a universal advantage in situations of conflict to be inalienably and manifestly rational in decision and motivation. Many of the attributes of rationality, as in several illustrations mentioned earlier, are strategic disabilities in certain conflict situations. It may be perfectly rational to wish oneself not altogether rational, or — if that language is philosophically objectionable — to wish for the power to suspend certain rational capabilities in particular situations. And one *can* suspend or destroy his own "rationality," at least to a limited extent; one can do this because the attributes that go to make up rationality are not inalienable, deeply personal, integral attributes of the human soul, but include such things as one's hearing aid, the reliability of the mails, the legal system, and the rationality of one's agents and partners. In principle, one might evade extortion equally well by drugging his brain, conspicuously isolating himself geographically, getting his assets legally impounded, or breaking the hand that he uses in signing checks. In a theory of strategy, several of these defenses can be represented as impairments of rationality if we wish to represent them so. A theory that makes rationality an explicit postulate is able not only to modify the postulate and examine its meaning but to take some of the mystery out of it. As a matter of fact, the paradoxical role of "rationality" in these conflict situations is evidence of the likely help that a systematic theory could provide.

And the results reached by a theoretical analysis of strategic behavior *are* often somewhat paradoxical; they often do contradict common sense or accepted rules. It is not true, as illustrated in the example of extortion, that in the face of a threat it is invariably an advantage to be rational, particularly if the fact of being rational or irrational cannot be concealed. It is not invariably an advantage, in the face of a threat, to have a communication system in good order, to have complete information, or to be in full command of one's own actions or of one's own assets. Mossadeq and my small children have already been referred to; but the same tactic is illustrated by the burning of bridges behind oneself to persuade an adversary that one cannot be induced to retreat. An old English law that made it a serious crime to *pay* tribute to coastal pirates does not necessarily appear either cruel or anomalous in the light of a theory of strategy. It is interesting that political democracy itself relies on a particular communication system in which the transmittal of authentic evidence is precluded: the mandatory secret ballot is a scheme to deny the voter any means of proving which way he voted. Being stripped of his power to prove how he voted, he is stripped of his power to be intimidated. Powerless to prove whether or not he complied with a threat, he knows — and so do those who would threaten him — that any punishment would be unrelated to the way he actually voted.

The well-known principle that one should pick good negotiators to represent him and then give them complete flexibility and authority — a principle commonly voiced by negotiators themselves — is by no means as self-evident as its proponents suggest; the power of a negotiator often rests on a manifest inability to make concessions and to meet demands.[8] Similarly, while prudence suggests leaving open a way of escape when one threatens an adversary with mutually painful reprisal, any visible means of escape may make the threat less credible. The very notion that it may be a strategic advantage to relinquish certain options deliberately, or even to give up all control over one's future actions and make his responses automatic, seems to be a hard one to swallow.

Many of these examples involve some denial of the value of skill, resourcefulness, rationality, knowledge, control, or freedom of choice. They are all, in principle, valid in certain circumstances; but seeing through their strangeness and comprehending the logic behind them is often a good deal easier if one has formalized the problem, studied it in the abstract, and identified analogies in other contexts where the strangeness is less of an obstacle to comprehension.

Another principle contrary to the usual first impression con-

[8] The administration of foreign aid presents numerous examples. See, for example, T. C. Schelling, "American Foreign Assistance," *World Politics* (July 1955), pp. 614–15.

cerns the relative virtues of clean and dirty bombs. Bernard Brodie has pointed out that when one considers the special requirements of deterrence, in contrast to the requirements of a war that one expects to fight, one may see some utility in the super-dirty bomb.[9] As remarked in Chapter 10, this conclusion is not so strange if we recognize the "balance of terror" as simply a massive modern version of an ancient institution, the exchange of hostages.

Here perhaps we perceive a disadvantage peculiar to civilized modern students of international affairs, by contrast with, say, Machiavelli or the ancient Chinese. We tend to identify peace, stability, and the quiescence of conflict with notions like trust, good faith, and mutual respect. To the extent that this point of view actually encourages trust and respect it is good. But where trust and good faith do not exist and cannot be made to by our acting as though they did, we may wish to solicit advice from the underworld, or from ancient despotisms, on how to make agreements work when trust and good faith are lacking and there is no legal recourse for breach of contract. The ancients exchanged hostages, drank wine from the same glass to demonstrate the absence of poison, met in public places to inhibit the massacre of one by the other, and even deliberately exchanged spies to facilitate transmittal of authentic information. It seems likely that a well-developed theory of strategy could throw light on the efficacy of some of those old devices, suggest the circumstances to which they apply, and discover modern equivalents that, though offensive to our taste, may be desperately needed in the regulation of conflict.

[9] Compare p. 239 below.

2

AN ESSAY ON BARGAINING

This chapter presents a tactical approach to the analysis of bargaining. The subject includes both explicit bargaining and the tacit kind in which adversaries watch and interpret each other's behavior, each aware that his own actions are being interpreted and anticipated, each acting with a view to the expectations that he creates. In economics the subject covers wage negotiations, tariff negotiations, competition where competitors are few, settlements out of court, and the real estate agent and his customer. Outside economics it ranges from the threat of massive retaliation to taking the right of way from a taxi.

Our concern will *not* be with the part of bargaining that consists of exploring for mutually profitable adjustments, and that might be called the "efficiency" aspect of bargaining. For example, can an insurance firm save money, and make a client happier, by offering a cash settlement rather than repairing the client's car; can an employer save money by granting a voluntary wage increase to employees who agree to take a substantial part of their wages in merchandise? Instead, we shall be concerned with what might be called the "distributional" aspect of bargaining: the situations in which a better bargain for one means less for the other. When the business is finally sold to the one interested buyer, what price does it go for? When two dynamite trucks meet on a road wide enough for one, who backs up?

These are situations that ultimately involve an element of pure bargaining — bargaining in which each party is guided mainly by his expectations of what the other will accept. But with each guided by expectations and knowing that the other is too, expectations become compounded. A bargain is struck when somebody makes a final, sufficient concession. Why does he concede?

Because he thinks the other will not. "I must concede because he won't. He won't because he thinks I will. He thinks I will because he thinks I think he thinks so. . . ." There is some range of alternative outcomes in which any point is better for both sides than no agreement at all. To insist on any such point is pure bargaining, since one always *would* take less rather than reach no agreement at all, and since one always *can* recede if retreat proves necessary to agreement. Yet if both parties are aware of the limits to this range, *any* outcome is a point from which at least one party would have been willing to retreat and the other knows it! There is no resting place.

There is, however, an outcome; and if we cannot find it in the logic of the situation we may find it in the tactics employed. The purpose of this chapter is to call attention to an important class of tactics, of a kind that is peculiarly appropriate to the logic of indeterminate situations. The essence of these tactics is some voluntary but irreversible sacrifice of freedom of choice. They rest on the paradox that the power to constrain an adversary may depend on the power to bind oneself; that, in bargaining, weakness is often strength, freedom may be freedom to capitulate, and to burn bridges behind one may suffice to undo an opponent.

BARGAINING POWER: THE POWER TO BIND ONESELF

"Bargaining power," "bargaining strength," "bargaining skill" suggest that the advantage goes to the powerful, the strong, or the skillful. It does, of course, if those qualities are defined to mean only that negotiations are won by those who win. But, if the terms imply that it is an advantage to be more intelligent or more skilled in debate, or to have more financial resources, more physical strength, more military potency, or more ability to withstand losses, then the term does a disservice. These qualities are by no means universal advantages in bargaining situations; they often have a contrary value.

The sophisticated negotiator may find it difficult to seem as obstinate as a truly obstinate man. If a man knocks at a door and says that he will stab himself on the porch unless given $10, he is more likely to get the $10 if his eyes are bloodshot. The threat of mutual destruction cannot be used to deter an adversary who is too unintelligent to comprehend it or too weak to enforce his will on those he represents. The government that cannot control its balance of payments, or collect taxes, or muster the political unity to defend itself, may enjoy assistance that would be denied it if it could control its own resources. And, to cite an example familiar from economic theory, "price leadership" in oligopoly may be an unprofitable distinction evaded by the small firms and assumed perforce by the large one.

Bargaining power has also been described as the power to fool and bluff, "the ability to set the best price for yourself and fool the other man into thinking this was your maximum offer."[1] Fooling and bluffing are certainly involved; but there are two kinds of fooling. One is deceiving about the facts; a buyer may lie about his income or misrepresent the size of his family. The other is purely tactical. Suppose each knows everything about the other, and each knows what the other knows. What is there to fool about? The buyer may say that, though he'd really pay up to twenty and the seller knows it, he is firmly resolved as a tactical matter not to budge above sixteen. If the seller capitulates, was he fooled? Or was he convinced of the truth? Or did the buyer really not know what he would do next if the tactic failed? If the buyer really "feels" himself firmly resolved, and bases his resolve on the conviction that the seller will capitulate, and the seller does, the buyer may say afterwards that he was "not fooling." Whatever has occurred, it is not adequately conveyed by the notions of bluffing and fooling.

How does one person make another believe something? The answer depends importantly on the factual question, "Is it true?" It is easier to prove the truth of something that is true than of something false. To prove the truth about our health we can call on a reputable doctor; to prove the truth about our costs or income we may let the person look at books that have been audited by a reputable firm or the Bureau of Internal Revenue. But to persuade him of something false we may have no such convincing evidence.

[1] J. N. Morgan, "Bilateral Monopoly and the Competitive Output," *Quarterly Journal of Economics*, 63:376n6 (August 1949).

When one wishes to persuade someone that he would not pay more than $16,000 for a house that is really worth $20,000 to him, what can he do to take advantage of the usually superior credibility of the truth over a false assertion? Answer: make it true. How can a buyer make it true? If he likes the house because it is near his business, he might move his business, persuading the seller that the house is really now worth only $16,000 to him. This would be unprofitable; he is no better off than if he had paid the higher price.

But suppose the buyer could make an irrevocable and enforceable bet with some third party, duly recorded and certified, according to which he would pay for the house no more than $16,000, or forfeit $5,000. The seller has lost; the buyer need simply present the truth. Unless the seller is enraged and withholds the house in sheer spite, the situation has been rigged against him; the "objective" situation — the buyer's true incentive — has been voluntarily, conspicuously, and irreversibly changed. The seller can take it or leave it. This example demonstrates that if the buyer can accept an irrevocable *commitment*, in a way that is unambiguously visible to the seller, he can squeeze the range of indeterminacy down to the point most favorable to him. It also suggests, by its artificiality, that the tactic is one that may or may not be available; whether the buyer can find an effective device for commiting himself may depend on who he is, who the seller is, where they live, and a number of legal and institutional arrangements (including, in our artificial example, whether bets are legally enforceable).

If both men live in a culture where "cross my heart" is universally accepted as potent, all the buyer has to do is allege that he will pay no more than $16,000, using this invocation of penalty, and he wins — or at least he wins if the seller does not beat him to it by shouting "$19,000, cross my heart." If the buyer is an agent authorized by a board of directors to buy at $16,000 but not a cent more, and the directors cannot constitutionally meet again for several months and the buyer cannot exceed his authority, and if all this can be made known to the seller, then the buyer "wins" — if, again, the seller has not tied himself up with a commitment to $19,000. Or, if the buyer can assert that he will pay no more than $16,000 so firmly that he would suffer intolerable loss of personal prestige or bargaining reputation by paying more, and if the fact of his paying more would necessarily be known, and if the seller appreciates all this, then a loud declaration by itself may provide the commitment. The device, of course, is a needless surrender of flexibility unless it can be made fully evident and understandable to the seller.

Incidentally, some of the more contractual kinds of commitments are not as effective as they at first seem. In the example of the self-inflicted penalty through the bet, it remains possible for the seller to seek out the third party and offer a modest sum in consideration of the latter's releasing the buyer from the bet, threatening to sell the house for $16,000 if the release is not forthcoming. The effect of the bet — as of most such contractual commitments — is to shift the locus and personnel of the negotiation, in the hope that the third party will be less available for negotiation or less subject to an incentive to concede. To put it differently, a *contractual* commitment is usually the assumption of a contingent "transfer cost," not a "real cost"; and if all interested parties can be brought into the negotiation the range of indeterminacy remains as it was. But if the third party were available only at substantial transportation cost, to that extent a truly irrevocable commitment would have been assumed. (If bets were made with a number of people, the "real costs" of bringing them into the negotiation might be made prohibitive.) [2]

[2] Perhaps the "ideal" solution to the bilateral monopoly problem is as follows. One member of the pair shifts his marginal cost curve so that joint profits are now zero at the output at which joint profits originally would have been maximized. He does this through an irrevocable sale-leaseback arrangement; he sells a royalty contract to some third party for a lump sum, the royalties so related to his output that joint costs exceed joint revenue at all other outputs. He cannot now afford to produce at any price or output except that price and output at which the entire original joint profits accrue to him; the other member of the bilateral monopoly sees the contract, appreciates the situation, and accepts his true minimum profits. The "winner" really gains the entire original profit via the lump sum for which he sold royalty rights; this profit does not affect his incentives because it is independent of what he produces. The third party pays the lump sum (minus a small discount for inducement) because he knows that the second party will have to capitulate and that therefore he will in fact get his contingent royalty. The hitch is that the royalty-rights buyer must not be available to the "losing member"; otherwise the latter can force him to renounce his royalty claim by threatening not to reach a bargain, thus restoring

The most interesting parts of our topic concern whether and how commitments can be taken; but it is worth while to consider briefly a model in which practical problems are absent — a world in which absolute commitments are freely available. Consider a culture in which "cross my heart" is universally recognized as absolutely binding. Any offer accompanied by this invocation is a final offer, and is so recognized. If each party knows the other's true reservation price, the object is to be first with a firm offer. Complete responsibility for the outcome then rests with the other, who can take it or leave it as he chooses (and who chooses to take it). Bargaining is all over; the commitment (that is, the first offer) wins.

Interpose some communication difficulty. They must bargain by letter; the invocation becomes effective when signed but cannot be known to the other until its arrival. Now when one party writes such a letter the other may already have signed his own, or may yet do so before the letter of the first arrives. There is then no sale; both are bound to incompatible positions. Each must now recognize this possibility of stalemate and take into account the likelihood that the other already has, or will have, signed his own commitment.

An asymmetry in communication may well favor the one who is (and is known to be) unavailable for the receipt of messages, for he is the one who cannot be deterred from his own commitment by receipt of the other's. (On the other hand, if the one who cannot communicate can feign ignorance of his own inability, the other too may be deterred from his own commitment by fear of the first's unwitting commitment.) If the commitments depend not just on words but on special forms or ceremonies, ignorance of the other party's commitment ceremonies may be an advantage if the ignorance is fully appreciated, since it makes the other aware that only his own restraint can avert stalemate.

Suppose only part of the population belongs to the cult in which "cross my heart" is (or is believed to be) absolutely bind-

---

the original marginal cost situation. But we may imagine the development of institutions that specialize in royalty purchases, whose ultimate success depends on a reputation for never renegotiating, and whose incentives can thus not be appealed to in any single negotiation.

ing. If everyone knows (and is known to know) everyone else's affiliation, those belonging to this particular cult have the advantage. They can commit themselves, the others cannot. If the buyer says "$16,000, cross my heart" his offer is final; if the seller says "$19,000" he is (and is known to be) only "bargaining."

If each does not know the other's true reservation price there is an initial stage in which each tries to discover the other's and misrepresent his own, as in ordinary bargaining. But the process of discovery and revelation becomes quickly merged with the process of creating and discovering commitments; the commitments permanently change, for all practical purposes, the "true" reservation prices. If one party has, and the other has not, the belief in a binding ceremony, the latter pursues the "ordinary" bargaining technique of *asserting* his reservation price, while the former proceeds to *make* his.

The foregoing discussion has tried to suggest both the plausibility and the logic of self-commitment. Some examples may suggest the relevance of the tactic, although an observer can seldom distinguish with confidence the consciously logical, the intuitive, or the inadvertent use of a visible tactic. First, it has not been uncommon for union officials to stir up excitement and determination on the part of the membership during or prior to a wage negotiation. If the union is going to insist on $2 and expects the management to counter with $1.60, an effort is made to persuade the membership not only that the management could pay $2 but even perhaps that the negotiators themselves are incompetent if they fail to obtain close to $2. The purpose — or, rather, a plausible purpose suggested by our analysis — is to make clear to the management that the negotiators could not accept less than $2 *even if they wished to* because they no longer control the members or because they would lose their own positions if they tried. In other words, the negotiators reduce the scope of their own authority and confront the management with the threat of a strike that the union itself cannot avert, even though it was the union's own action that eliminated its power to prevent the strike.

Something similar occurs when the United States Government

negotiates with other governments on, say, the uses to which foreign assistance will be put, or tariff reduction. If the executive branch is free to negotiate the best arrangement it can, it may be unable to make any position stick and may end by conceding controversial points because its partners know, or believe obstinately, that the United States would rather concede than terminate the negotiations. But, if the executive branch negotiates under legislative authority, with its position constrained by law, and it is evident that Congress will not be reconvened to change the law within the necessary time period, then the executive branch has a firm position that is visible to its negotiating partners.

When national representatives go to international negotiations knowing that there is a wide range of potential agreement within which the outcome will depend on bargaining, they seem often to create a bargaining position by public statements, statements calculated to arouse a public opinion that permits no concessions to be made. If a binding public opinion can be cultivated and made evident to the other side, the initial position can thereby be made visibly "final."

These examples have certain characteristics in common. First, they clearly depend not only on incurring a commitment but on communicating it persuasively to the other party. Second, it is by no means easy to establish the commitment, nor is it entirely clear to either of the parties concerned just how strong the commitment is. Third, similar activity may be available to the parties on both sides. Fourth, the possibility of commitment, though perhaps available to both sides, is by no means equally available; the ability of a democratic government to get itself tied by public opinion may be different from the ability of a totalitarian government to incur such a commitment. Fifth, they all run the risk of establishing an immovable position that goes beyond the ability of the other to concede, and thereby provoke the likelihood of stalemate or breakdown.

INSTITUTIONAL AND STRUCTURAL CHARACTERISTICS
OF THE NEGOTIATION

Some institutional and structural characteristics of bargaining situations may make the commitment tactic easy or difficult to use, or make it more available to one party than the other, or affect the likelihood of simultaneous commitment or stalemate.

Use of a Bargaining Agent. The use of a bargaining agent affects the power of commitment in at least two ways. First, the agent may be given instructions that are difficult or impossible to change, such instructions (and their inflexibility) being visible to the opposite party. The principle applies in distinguishing the legislative from the executive branch, or the management from the board of directors, as well as to a messenger-carried offer when the bargaining process has a time limit and the principal has interposed sufficient distance between himself and his messenger to make further communication evidently impossible before the time runs out.

Second, an "agent" may be brought in as a principal in his own right, with an incentive structure of his own that differs from his principal's. This device is involved in automobile insurance; the private citizen, in settling out of court, cannot threaten suit as effectively as the insurance company since the latter is more conspicuously obliged to carry out such threats to maintain its own reputation for subsequent accidents.[3]

Secrecy vs. Publicity. A potent means of commitment, and sometimes the only means, is the pledge of one's reputation. If national representatives can arrange to be charged with appeasement for every small concession, they place concession visibly beyond their own reach. If a union with other plants to deal with can arrange to make any retreat dramatically visible, it places its bargaining reputation in jeopardy and thereby becomes visibly incapable of serious compromise. (The same convenient jeopardy is the basis for the universally exploited defense, "If I did it for you I'd have to do it for everyone else.") But to commit in this

---

[3] The formal solution to the right-of-way problem in automobile traffic may be that the winner is the one who first becomes fully and visibly insured against all contingencies; since he then has no incentive to avoid accident, the other must yield and knows it. (The latter cannot counter in kind; no company will insure him now that the first is insured.) More seriously, the pooling of strike funds among unions reduces the visible incentive on each individual union to avoid a strike. As in the bilateral monopoly solution suggested earlier, there is a transfer of interest to a third party with a resulting visible shift in one's own incentive structure.

fashion publicity is required. Both the initial offer and the final outcome would have to be known; and if secrecy surrounds either point, or if the outcome is inherently not observable, the device is unavailable. If one party has a "public" and the other has not, the latter may try to neutralize his disadvantage by excluding the relevant public; or if both parties fear the potentialities for stalemate in the simultaneous use of this tactic, they may try to enforce an agreement on secrecy.

Intersecting Negotiations. If a union is simultaneously engaged, or will shortly be engaged, in many negotiations while the management has no other plants and deals with no other unions, the management cannot convincingly stake its bargaining reputation while the union can. The advantage goes to the party that can persuasively point to an array of other negotiations in which its own position would be prejudiced if it made a concession in this one. (The "reputation value" of the bargain may be less related to the outcome than to the firmness with which some initial bargaining position is adhered to.) Defense against this tactic may involve, among other things, both misinterpretation of the other party's position and an effort to make the eventual outcome incommensurable with the initial positions. If the subjects under negotiation can be enlarged in the process of negotiation, or the wage figure replaced by fringe benefits that cannot be reduced to a wage equivalent, an "out" is provided to the party that has committed itself; and the availability of this "out" weakens the commitment itself, to the disadvantage of the committed party.

Continuous Negotiations. A special case of interrelated negotiations occurs when the same two parties are to negotiate other topics, simultaneously or in the future. The logic of this case is more subtle; to persuade the other than one cannot afford to recede, one says in effect, "If I conceded to you here, you would revise your estimate of me in our other negotiations; to protect my reputation with you I must stand firm." The second party is simultaneously the "third party" to whom one's bargaining reputation can be pledged. This situation occurs in the threat of local resistance to local aggression. The party threatening achieves its commitment, and hence the credibility of its threat, not by referring to what it would gain from carrying out the threat in this particular instance but by pointing to the long-run value of a fulfilled threat in enhancing the credibility of future threats.

The Restrictive Agenda. When there are two objects to negotiate, the decision to negotiate them simultaneously or in separate forums or at separate times is by no means neutral to the outcome, particularly when there is a latent extortionate threat that can be exploited only if it can be attached to some more ordinary, legitimate, bargaining situation. The protection against extortion depends on refusal, unavailability, or inability, to negotiate. But if the object of the extortionate threat can be brought onto the agenda with the other topic, the latent threat becomes effective.

Tariff bargaining is an example. If reciprocal tariffs on cheese and automobiles are to be negotiated, one party may alter the outcome by threatening a purely punitive change in some other tariff. But if the bargaining representatives of the threatened party are confined to the cheese-automobile agenda, and have no instructions that permit them even to take cognizance of other commodities, or if there are ground rules that forbid mention of other tariffs while cheese and automobiles remain unsettled, this extortionate weapon must await another opportunity. If the threat that would be brought to the conference table is one that cannot stand publicity, publicity itself may prevent its effective communication.

The Possibility of Compensation. As Fellner has pointed out, agreement may be dependent on some means of redistributing costs or gains.[4] If duopolists, for example, divide markets in a way that maximizes their combined profits, some initial accrual of profits is thereby determined; any other division of the profits requires that one firm be able to compensate the other. If the fact of compensation would be evidence of illegal collusion, or if the motive for compensation would be misunderstood by the stockholders, or if the two do not sufficiently trust each other, some less optimum level of *joint* profits may be required in order that the

---

[4] W. Fellner, *Competition Among the Few* (New York, 1949), pp. 34–35, 191–97, 231–32, 234.

initial accrual of profits to the two firms be in closer accordance with an agreed division of gains between them.

When agreement must be reached on something that is inherently a one-man act, any division of the cost depends on compensation. The "agenda" assumes particular importance in these cases, since a principal means of compensation is a concession on some other object. If two simultaneous negotiations can be brought into a contingent relationship with each other, a means of compensation is available. If they are kept separate, each remains an indivisible object.

It may be to the advantage of one party to keep a bargain isolated, and to the other to join it to some second bargain. If there are two projects, each with a cost of three, and each with a value of two to A and a value of four to B, and each is inherently a "one-man" project in its execution, and if compensation is institutionally impossible, B will be forced to pay the entire cost of each as long as the two projects are kept separate. He cannot usefully threaten nonperformance, since A has no incentive to carry out either project by himself. But if B can link the projects together, offering to carry out one while A carries out the other, and can effectively threaten to abandon both unless A carries out one of them, A is left an option with a gain of four and a cost of three, which he takes, and B cuts his cost in half.

An important limitation of economic problems, as prototypes of bargaining situations, is that they tend disproportionately to involve divisible objects and compensable activities. If a drainage ditch in the back of one house will protect both houses; and if it costs $1,000 and is worth $800 to each home-owner; neither would undertake it separately, but we nevertheless usually assume that they will get together and see that this project worth $1,600 to the two of them gets carried out. But if it costs 10 hours a week to be scoutmaster, and each considers it worth 8 hours of his time to have a scout troop but one man must do the whole job, it is far from certain that the neighbors will reach a deal according to which one puts 10 hours on the job and the other pays him cash or does 5 hours' gardening for him. When two cars meet on a narrow road, the ensuing deadlock is aggravated by the absence of a custom of bidding to pay for the right of way. Parliamentary dead-locks occur when logrolling is impracticable. Measures that re-quire unanimous agreement can often be initiated only if several are bundled together.[5]

The Mechanics of Negotiation. A number of other characteristics deserve mention, although we shall not work out their implica-tions. Is there a penalty on the conveyance of false information? Is there a penalty on called bluffs, that is, can one put forth an offer and withdraw it after it has been accepted? Is there a penalty on hiring an agent who pretends to be an interested party and makes insincere offers, simply to test the position of the other party? Can all interested parties be recognized? Is there a time limit on the bargaining? Does the bargaining take the particular structure of an auction, a Dutch auction, a sealed bid system, or some other formal arrangement? Is there *a status quo,* so that unavailability for negotiation can win the *status quo* for the party that prefers it? Is renegotiation possible in case of stalemate? What are the costs of stalemate? Can compliance with the agree-ment be observed? What, in general, are the means of communi-cation, and are any of them susceptible of being put out of order by one party or the other? If there are several items to negotiate, are they negotiated in one comprehensive negotiation, separately in a particular order so that each piece is finished before the next is taken up, or simultaneously through different agents or under different rules.

The importance of many of these structural questions becomes evident when one reflects on parliamentary technique. Rules that permit a president to veto an appropriation bill only in its en-tirety, or that require each amendment to be voted before the original act is voted on, or a priority system accorded to different kinds of motions, substantially alter the incentives that are brought to bear on each action. One who might be pressured into choosing second best is relieved of his vulnerability if he can vote earlier to eliminate that possibility, thereby leaving only

---

[5] Inclusion of a provision on the Saar in the "Paris Agreements" that ended the occupation of Western Germany may have reflected either this principle or the one in the preceding paragraph.

first and third choices about which his preference is known to be so strong that no threat will be made.

Principles and Precedents. To be convincing, commitments usually have to be qualitative rather than quantitative, and to rest on some rationale. It may be difficult to conceive of a really firm commitment to $2.07½; why not $2.02¼? The numerical scale is too continuous to provide good resting places, except at nice round numbers like $2.00. But a commitment to the *principle* of "profit sharing," "cost-of-living increases," or any other basis for a numerical calculation that comes out at $2.07½, may provide a foothold for a commitment. Furthermore, one may create something of a commitment by putting the principles and precedents themselves in jeopardy. If in the past one has successfully maintained the principle of, say, nonrecognition of governments imposed by force, and elects to nail his demands to that principle in the present negotiation, he not only adduces precedent behind his claim but risks the principle itself. Having pledged it, he may persuade his adversary that he would accept stalemate rather than capitulate and discredit the principle.

Casuistry. If one reaches the point where concession is advisable, he has to recognize two effects: it puts him closer to his opponent's position, and it affects his opponent's estimate of his firmness. Concession not only may be construed as capitulation, it may mark a prior commitment as a fraud, and make the adversary skeptical of any new pretense at commitment. One, therefore, needs an "excuse" for accommodating his opponent, preferably a rationalized reinterpretation of the original commitment, one that is persuasive to the adversary himself.

More interesting is the use of casuistry to release an opponent from a commitment. If one can demonstrate to an opponent that the latter is not committed, or that he has miscalculated his commitment, one may in fact undo or revise the opponent's commitment. Or if one can confuse the opponent's commitment, so that his constituents or principals or audience cannot exactly identify compliance with the commitment — show that "productivity" is ambiguous, or that "proportionate contributions" has several meanings — one may undo it or lower its value. In these cases it is to the opponent's disadvantage that this commitment be successfully refuted by argument. But when the opponent has resolved to make a moderate concession one may help him by proving that he *can* make a moderate concession consistent with his former position, and that if he does there are no grounds for believing it to reflect on his original principles. One must seek, in other words, a rationalization by which to deny oneself too great a reward from the opponent's concession, otherwise the concession will not be made.[6]

THE THREAT

When one threatens to fight if attacked or to cut his price if his competitor does, the threat is no more than a communication of one's own incentives, designed to impress on the other the automatic consequences of his act. And, incidentally, if it succeeds in deterring, it benefits both parties.

But more than communication is involved when one threatens an act that he would have no incentive to perform but that is designed to deter through its promise of mutual harm. To threaten massive retaliation against small encroachments is of this nature, as is the threat to bump a car that does not yield the right of way or to call a costly strike if the wage is not raised a few cents. The distinctive feature of this threat is that the threatener has no in-

[6] In many textbook problems, such as bilateral monopoly between firms, the ends of the bargaining range are points of zero profits for one or the other party; and to settle for one's minimum position is no better than no settlement at all. But, apart from certain buying and selling situations, there are commonly limits on the range of acceptable outcomes, and the least favorable outcome that one is free to accept may be substantially superior to stalemate. In these cases one's overriding purpose may be to forestall any misguided commitment by the other party. If the truth is more demonstrable than a false position, a conservative initial position is indicated, as it is if any withdrawal from an initial "advanced" position would discredit any subsequent attempt to convey the truth. Actually, though a person does not commonly invite penalties on his own behavior, the existence of an enforceable penalty on falsehood would be of assistance; if one can demonstrate, for example, his cost or income position by showing his income tax return, the penalties on fraud may enhance the value of this evidence.

Even the "pure" bilateral monopoly case becomes somewhat of this nature if the bargaining is conducted by agents or employees whose rewards are more dependent on *whether* agreement is reached than on how favorable the terms of the agreement are.

centive to carry it out either before the event or after. He does have an incentive to bind himself to fulfill the threat, if he thinks the threat may be successful, because the threat and not its fulfillment gains the end; and fulfillment is not required if the threat succeeds. The more certain the contingent fulfillment is, the less likely is actual fulfillment. But the threat's efficacy depends on the credulity of the other party, and the threat is ineffectual unless the threatener can rearrange or display his own incentives so as to demonstrate that he would, *ex post*, have an incentive to carry it out.[7]

We are back again at the commitment. How can one commit himself in advance to an act that he would in fact prefer not to carry out in the event, in order that his commitment may deter the other party? One can of course bluff, to persuade the other falsely that the costs or damages to the threatener would be minor or negative. More interesting, the one making the threat may pretend that he himself erroneously believes his own costs to be small, and therefore would mistakenly go ahead and fulfill the threat. Or perhaps he can pretend a revenge motivation so strong as to overcome the prospect of self-damage; but this option is probably most readily available to the truly revengeful. Otherwise he must find a way to commit himself.

One may try to stake his reputation on fulfillment, in a manner that impresses the threatened person. One may even stake his reputation *with the threatened person himself*, on grounds that it would be worth the costs and pains to give a lesson to the latter if he fails to heed the threat. Or one may try to arrange a legal commitment, perhaps through contracting with a third party.[8]

---

[7] Incidentally, the deterrent threat has some interesting quantitative characteristics, reflecting the general asymmetry between rewards and punishments. It is not necessary, for example, that the threat promise more damage to the party threatened than to the party carrying it out. The threat to smash an old car with a new one may succeed if believed, or to sue expensively for small damages, or to start a price war. Also, as far as the power to deter is concerned, there is no such thing as "too large" a threat; if it is large enough to succeed, it is not carried out anyway. A threat is only "too large" if its very size interferes with its credibility. Atomic destruction for small misdemeanors, like expensive incarceration for overtime parking, would be superfluous but not exorbitant unless the threatened person considered it too awful to be real and ignored it.

[8] Mutual defense treaties among strong and weak nations might best be

Or if one can turn the whole business over to an agent whose salary (or business reputation) depends on carrying out the threat but who is unalterably relieved of any responsibility for the further costs, one may shift the incentive.

The commitment problem is nicely illustrated by the legal doctrine of the "last clear chance" which recognizes that, in the events that led up to an accident, there was some point at which the accident became inevitable as a result of prior actions, and that the abilities of the two parties to prevent it may not have expired at the same time. In bargaining, the commitment is a device to leave the last clear chance to decide the outcome with the other party, in a manner that he fully appreciates; it is to relinquish further initiative, having rigged the incentives so that the other party must choose in one's favor. If one driver speeds up so that he cannot stop, and the other realizes it, the latter has to yield. A legislative rider at the end of a session leaves the President the last clear chance to pass the bill. This doctrine helps to understand some of those cases in which bargaining "strength" inheres in what is weakness by other standards. When a person — or a country — has lost the power to help himself, or the power to avert mutual damage, the other interested party has no choice but to assume the cost or responsibility. "Coercive deficiency" is the term Arthur Smithies uses to describe the tactic of deliberately exhausting one's annual budgetary allowance so early in the year that the need for more funds is irresistibly urgent.[9]

A related tactic is maneuvering into a *status quo* from which one can be dislodged only by an overt act, an act that precipitates mutual damage because the maneuvering party has relinquished the power to retreat. If one carries explosives visibly on his person, in a manner that makes destruction obviously inevitable for himself and for any assailant, he may deter assault much more than if he retained any control over the explosives. If one com-

---

viewed in this light, that is, not as undertaken to reassure the small nations nor in exchange for a *quid pro quo*, but rather as a device for surrendering an embarrassing freedom of choice.

[9] A. Smithies, *The Budgetary Process in the United States* (New York, 1955), pp. 40, 56. One solution is the short tether of an apportionment process. See also T. C. Schelling, "American Foreign Assistance," *World Politics*, 7:609–625 (July 1955), regarding the same principle in foreign aid allocations.

mits a token force of troops that would be unable to escape, the commitment to full resistance is increased. Walter Lippmann has used the analogy of the plate glass window that helps to protect a jewelry store: anyone can break it easily enough, but not without creating an uproar.

Similar techniques may be available to the one threatened. His best defense, of course, is to carry out the act before the threat is made; in that case there is neither incentive nor commitment for retaliation. If he cannot hasten the act itself, he may commit himself to it; if the person to be threatened is already committed, the one who would threaten cannot deter with his threat, he can only make certain the mutually disastrous consequences that he threatens.[10] If the person to be threatened can arrange before the threat is made to share the risk with others (as suggested by the insurance solution to the right-of-way problem mentioned earlier) he may become so visibly unsusceptible to the threat as to dissuade the threatener. Or if by any other means he can either change or misrepresent his own incentives, to make it appear that he would gain in spite of threat fulfillment (or perhaps only that he thinks he would), the threatener may have to give up the threat as costly and fruitless; or if one can misrepresent himself as either unable to comprehend a threat, or too obstinate to heed it, he may deter the threat itself. Best of all may be *genuine* ignorance, obstinacy, or simple disbelief, since it may be more convincing to the prospective threatener; but of course if it fails to persuade him and he commits himself to the threat, both sides lose. Finally, both the threat and the commitment have to be communicated; if the threatened person can be unavailable for messages, or can destroy the communication channels, even though

[10] The system of supplying the police with traffic tickets that are numbered and incapable of erasures makes it possible for the officer, by writing in the license number of the car before speaking to the driver, to preclude the latter's threat. Some trucks carry signs that say, "Alarm and lock system not subject to the driver's control." The time lock on bank vaults serves much the same purpose, as does the mandatory secret ballot in elections. So does starting an invasion with a small advance force that, though too small and premature to win the objective, attaches too much "face" to the enterprise to permit withdrawal: the larger force can then be readied without fear of inviting a purely deterrent threat. At many universities the faculty is protected by a rule that denies instructors the power to change a course grade once it has been recorded.

he does so in an obvious effort to avert threat, he may deter the threat itself.[11] But the time to show disbelief or obstinacy is before the threat is made, that is, before the commitment is taken, not just before the threat is fulfilled; it does no good to be incredulous, or out of town, when the messenger arrives with the committed threat.

In threat situations, as in ordinary bargaining, commitments are not altogether clear; each party cannot exactly estimate the costs and values to the other side of the two related actions involved in the threat; the process of commitment may be a progressive one, the commitments acquiring their firmness by a sequence of actions. Communication is often neither entirely impossible nor entirely reliable; while certain evidence of one's commitment can be communicated directly, other evidence must travel by newspaper or hearsay, or be demonstrated by actions. In these cases the unhappy possibility of both acts occurring, as a result of simultaneous commitment, is increased. Furthermore, the recognition of this possibility of simultaneous commitment becomes itself a deterrent to the taking of commitments.[12]

In case a threat is made and fails to deter, there is a second stage prior to fulfillment in which *both* parties have an interest in undoing the commitment. The purpose of the threat is gone, its deterrence value is zero, and only the commitment exists to motivate fulfillment. This feature has, of course, an analogy with stalemate in ordinary bargaining, stalemate resulting from both parties' getting committed to incompatible positions, or one party's

[11] The racketeer cannot sell protection if he cannot find his customer at home; nor can the kidnapper expect any ransom if he cannot communicate with friends or relatives. Thus, as a perhaps impractical suggestion, a law that required the immediate confinement of all interested friends and relatives when a kidnapping occurred might make the prospects for ransom unprofitably dim. The rotation of watchmen and policemen, or their assignment in random pairs, not only limits their exploitation of bribes but protects them from threats.

[12] It is a remarkable institutional fact that there is no simple, universal way for persons or nations to assume commitments of the kind we have been discussing. There are numerous ways they can try, but most of them are quite ambiguous, unsure, or only occasionally available. In the "cross-my-heart" society adverted to earlier, bargaining theory would reduce itself to game strategy and the mechanics of communication; but in most of the contemporary world the topic is mainly an empirical and institutional one of who can commit, how, and with what assurance of appreciation by the other side.

mistakenly committing himself to a position that the other truly would not accept. If there appears a possibility of undoing the commitment, *both* parties have an interest in doing so. How to undo it is a matter on which their interests diverge, since different ways of undoing it lead to different outcomes. Furthermore, "undoing" does not mean neglecting a commitment regardless of reputation; "undoing," if the commitment of reputation was real, means disconnecting the threat from one's reputation, perhaps one's own reputation with the threatened person himself. It is therefore a subtle and tenuous situation in which, though both have an interest in undoing the commitment, they may be quite unable to collaborate in undoing it.

Special care may be needed in defining the threat, both the act that is threatened against and the counter act that is threatened. The difficulty arises from the fact, just noted, that once the former has been done the incentive to perform the later has disappeared. The credibility of the threat before the act depends on how visible to the threatened party is the inability of the threatening party to rationalize his way out of his commitment once it has failed its purpose. Any loopholes the threatening party leaves himself, if they are visible to the threatened party, weaken the visible commitment and hence reduce the credibility of the threat. (An example may be the ambiguous treatment of Quemoy in the Formosa Resolution and Treaty.)

It is essential, therefore, for maximum credibility, to leave as little room as possible for judgment or discretion in carrying out the threat. If one is committed to punish a certain type of behavior when it reaches certain limits, but the limits are not carefully and objectively defined, the party threatened will realize that when the time comes to decide whether the threat must be enforced or not, his interest and that of the threatening party will coincide in an attempt to avoid the mutually unpleasant consequences.

In order to make a threat precise, so that its terms are visible both to the threatened party and to any third parties whose reaction to the whole affair is of value to the adversaries, it may be necessary to introduce some arbitrary elements. The threat must involve overt acts rather than intentions; it must be attached to the visible deeds, not invisible ones; it may have to attach itself to certain ancillary actions that are of no consequence in themselves to the threatening party. It may, for example, have to put a penalty on the carrying of weapons rather than their use; on suspicious behavior rather than observed misdemeanors; on proximity to a crime rather than the crime itself. And, finally, the act of punishment must be one whose effect or influence is clearly discernible.[13]

In order that one be able to pledge his reputation behind a threat, there must be continuity between the present and subsequent issues that will arise. This need for continuity suggests a means of making the original threat more effective; if it can be decomposed into a series of consecutive smaller threats, there is an opportunity to demonstrate on the first few transgressions that the threat will be carried out on the rest. Even the first few become more plausible, since there is a more obvious incentive to fulfill them as a "lesson."

This principle is perhaps most relevant to acts that are inherently a matter of degree. In foreign aid programs the overt act of terminating assistance may be so obviously painful to both sides as not to be taken seriously by the recipient, but if each small misuse of funds is to be accompanied by a small reduction in assistance, never so large as to leave the recipient helpless nor to provoke a diplomatic breach, the willingness to carry it out will receive more credulity; or if it does not at first, a few lessons may be persuasive without too much damage.[14]

The threatening party may not, of course, be able to divide the act into steps. (Both the act to be deterred and the punishment must be divisible.) But the principle at least suggests the unwisdom of defining aggression, or transgression, in terms of some

[13] During 1950, the Economic Cooperation Administration declared its intention to reward Marshall Plan countries that followed especially sound policies, and to penalize those that did not, through the device of larger or smaller aid allotments. But since the base figures had not been determined, and since their determination would ultimately involve judgment rather than formulas, there would be no way afterwards to see whether in fact the additions and subtractions were made, and the plan suffered from implausibility.

[14] Perhaps the common requirement for amortization of loans at frequent intervals, rather than in a lump sum at the end of the loan period, reflects an analogous principle, as does the custom of giving frequent examinations in a college course to avoid letting a student's failure hinge exclusively on a single grading decision after the course is finished.

critical degree or amount that will be deemed intolerable. When the act to be deterred is inherently a sequence of steps whose cumulative effect is what matters, a threat geared to the increments may be more credible than one that must be carried out either all at once or not at all when some particular point has been reached. It may even be impossible to define a "critical point" with sufficient clarity to be persuasive.

To make the threatened acts divisible, the acts themselves may have to be modified. Parts of an act that cannot be decomposed may have to be left out; ancillary acts that go with the event, though of no interest in themselves, may be objects to which a threat can effectively be attached. For example, actions that are only preparatory to the main act, and by themselves do no damage, may be susceptible of chronological division and thus be effective objects of the threat. The man who would kick a dog should be threatened with modest punishment for each step toward the dog, even though his proximity is of no interest in itself.

Similar to decomposing a threat into a series is starting a threat with a punitive act that grows in severity with the passage of time. Where a threat of death by violence might not be credited, cutting off the food supply might bring submission. For moral or public relations purposes, this device may in fact leave the "last clear chance" to the other, whose demise is then blamed on his stubbornness if the threat fails. But in any case the threatener gets his overt act out of the way while it is still preliminary and minor, rather than letting it stand as a final, dreadful, and visible obstacle to his resolution. And if the suffering party is the only one in a position to know, from moment to moment, how near to catastrophe they have progressed, his is the last clear chance in a real sense. Furthermore, the threatener may be embarrassed by his adversary's collapse but not by his discomfort; and the device may therefore transform a dangerous once-for-all threat into a less costly continuous one. Tenants are less easily removed by threat of forcible eviction than by simply shutting off the utilities.[15]

---

[15] This seems to be the tactic that avoided an explosion and induced de Gaulle's forces to vacate a province they had occupied in Northern Italy in June 1945, after they had announced that any effort of their allies to dislodge them would

A piecemeal approach may also be used by the threatened person. If he cannot obviate the threat by hastening the entire act, he may hasten some initial stage that clearly commits him to eventual completion. Or, if his act is divisible while the threatener's retaliation comes only in the large economy size, performing it as a series of increments may deny the threatener the dramatic overt act that would trigger his response.

THE PROMISE

Among the legal privileges of corporations, two that are mentioned in textbooks are the right to sue and the "right" to be sued. Who wants to be sued! But the right to be sued is the power to make a promise: to borrow money, to enter a contract, to do business with someone who might be damaged. If suit does arise, the "right" seems a liability in retrospect; beforehand it was a prerequisite to doing business.

In brief, the right to be sued is the power to accept a commitment. In the commitments discussed up to this point, it was essential that one's adversary (or "partner," however we wish to describe him) not have the power to release one from the commitment; the commitment was, in effect, to some third party, real or fictitious. The promise is a commitment to the second party in the bargain and is required whenever the final action of one or of each is outside the other's control. It is required whenever an agreement leaves any incentive to cheat.[16]

This need for promises is more than incidental; it has an institutional importance of its own. It is not always easy to make a convincing, self-binding, promise. Both the kidnapper who would like to release his prisoner, and the prisoner, may search desperately for a way to commit the latter against informing on his captor, without finding one. If the victim has committed an act

---

be treated as a hostile act. See Harry S Truman, *Year of Decisions* (New York, 1955), pp. 239–42; and Winston S. Churchill, *Triumph and Tragedy*, vol. VI of *The Second World War* (Boston, 1953), pp. 566–68.

[16] The threat may seem to be a promise if the pledge behind it is only one's reputation with his adversary; but it is not a promise from which the second party can unilaterally release the threatener, since he cannot convincingly dissociate his own future estimate of the threatener from the latter's performance.

whose disclosure could lead to blackmail, he may confess it; if not, he might commit one in the presence of his captor, to create the bond that will ensure his silence. But these extreme possibilities illustrate how difficult, as well as important, it may be to assume a promise. If the law will not enforce price agreements; or if the union is unable to obligate itself to a no-strike pledge; or if a contractor has no assets to pay damages if he loses a suit, and the law will not imprison debtors; or if there is no "audience" to which one can pledge his reputation; it may not be possible to strike a bargain, or at least the same bargain that would otherwise be struck.

Bargaining may have to concern itself with an "incentive" system as well as the division of gains. Oligopolists may lobby for a "fair-trade" law; or exchange shares of stocks. An agreement to stay out of each other's market may require an agreement to redesign the products to be unsuitable in each other's area. Two countries that wish to agree not to make military use of an island may have to destroy the usefulness of the island itself. (In effect, a "third-party commitment" has to be assumed when an effective "second-party commitment" cannot be devised.) [17]

Fulfillment is not always observable. If one sells his vote in a secret election, or a government agrees to recommend an act to its parliament, or an employee agrees not to steal from inventory, or a teacher agrees to keep his political opinions out of class, or a country agrees to stimulate exports "as much as possible," there is no reliable way to observe or measure compliance. The observable outcome is subject to a number of influences, only one of which is covered by the agreement. The bargain may therefore have to be expressed in terms of something observable, even though what is observable is not the intended object of the bargain. One may have to pay the bribed voter if the election is won, not on how he voted; to pay a salesman a commission on sales, rather than on skill and effort; to reward policemen according to statistics on crime rather than on attention to duty; or to punish all employees for the transgressions of one. And, where performance is a matter of degree, the bargain may have to define arbitrary limits distinguishing performance from nonperformance; a speci-

[17] In an earlier age, hostages were exchanged.

fied loss of inventory treated as evidence of theft; a specified increase in exports considered an "adequate" effort; specified samples of performance taken as representative of total performance.[18]

The tactic of decomposition applies to promises as well as to threats. What makes many agreements enforceable is only the recognition of future opportunities for agreement that will be eliminated if mutual trust is not created and maintained, and whose value outweighs the momentary gain from cheating in the present instance. Each party must be confident that the other will not jeopardize future opportunities by destroying trust at the outset. This confidence does not always exist; and one of the purposes of piecemeal bargains is to cultivate the necessary mutual expectations. Neither may be willing to trust the other's prudence (or the other's confidence in the first's prudence, and so forth) on a large issue. But, if a number of preparatory bargains can be struck on a small scale, each may be willing to risk a small investment to create a tradition of trust. The purpose is to let each party demonstrate that he appreciates the need for trust and that he knows the other does too. So, if a major issue has to be negotiated, it may be necessary to seek out and negotiate some minor items for "practice," to establish the necessary confidence in each other's awareness of the long-term value of good faith.

Even if the future will bring no recurrence, it may be possible to create the equivalence of continuity by dividing the bargaining issue into consecutive parts. If each party agrees to send a million dollars to the Red Cross on condition the other does, each may be tempted to cheat if the other contributes first, and each one's anticipation of the other's cheating will inhibit agreement. But if the contribution is divided into consecutive small contributions, each can try the other's good faith for a small price. Furthermore, since each can keep the other on short tether to the finish, no one ever need risk more than one small contribution at a time. Finally, this change in the incentive structure itself takes most

[18] Inability to assume an enforceable promise, like inability to perform the activity demanded, may protect one from an extortionate threat. The mandatory secret ballot is a nuisance to the voter who would like to sell his vote, but protection to the one who would fear coercion.

of the risk out of the initial contribution; the value of established trust is made obviously visible to both.

Preparatory bargains serve another purpose. Bargaining can only occur when at least one party takes initiative in proposing a bargain. A deterrent to initiative is the information it yields, or may seem to yield, about one's eagerness. But if each has visible reason to expect the other to meet him half way, because of a history of successful bargaining, that very history provides protection against the inference of overeagerness.[19]

AN ILLUSTRATIVE GAME

Various bargaining situations involving commitments, threats, promises, and communication problems, can be illustrated by variants of a game in which each of two persons has a pair of alternatives from which to choose. North chooses either A or a; East chooses either B or β. Each person's gain depends on the choices of both. Each of the four possible combined choices, AB, $A\beta$, aB, or $a\beta$, yields a particular gain or loss for North and a particular gain or loss for East. No compensation is payable between North and East. In general, each person's preference may depend on the choice the other makes.

Each such game can be quantitatively represented in a two-dimensional graph, with North's gain measured vertically and East's horizontally, and the values of the four combined choices denoted by points labeled AB, $A\beta$, $a\beta$, and aB. In spite of the simplicity of the game there is actually a large number of qualitatively different variants, depending not only on the relative positions of the four points in the plane but also on the "rules" about order of moves, possibility of communication, availability of means of commitment, enforceability of promises, and whether

---

[19] Perhaps two adversaries who look forward to some large negotiated settlement would do well to keep avenues open for negotiation of minor issues. If, for example, the number of loose ends in dispute between East and West should narrow down so much that nothing remains to be negotiated but the "ultimate issue" (some final, permanent disposition of all territories and armaments) the possibility of even opening negotiations on the latter might be jeopardized. Or, if the minor issues are not disposed of, but become so attached to the "big" issue that willingness to negotiate on them would be construed as overeagerness on the whole settlement, the possibility of preparatory bargains might disappear.

two or more games between two persons can be joined together. The variations can be multiplied almost without limit by selecting different hypotheses about what each player knows or guesses about the "values" of the four outcomes for the other player, and what he guesses the other party guesses about himself. For convenience we assume here that the eight "values" are obvious in an obvious way to both persons. And, just as we have ruled out compensation, we rule out also threats of actions that lie outside the game. A very small sample of such games is presented.

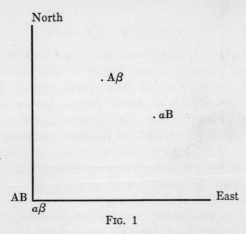

FIG. 1

Figure 1 represents an "ordinary" bargaining situation if we adopt the rule that North and East must reach explicit agreement before they choose. $A\beta$ and αB can be thought of as alternative agreements that they may reach, while AB and $\alpha\beta$, with zero values for both persons, can be interpreted as the bargaining equivalent of "no sale." Whoever can first commit himself wins. If North can commit himself to A he will secure $A\beta$, since he leaves East a choice between $A\beta$ and AB and the former is obviously East's choice under the circumstances. If East could have committed himself first to B, however, North would have been restricted to a choice of αB or no agreement (that is, of αB or AB) and would have agreed to αB. As a matter of fact, first commitment is a kind of "first move"; and in a game with the same numbers but with moves in turn, first move would be an advantage. If, by mistake, both parties get committed, North to A and East to B, they lock themselves in stalemate at AB.

Figure 2 illustrates a deterrent threat if we interpret AB as the *status quo*, with North planning a shift to α (leading to αB) and East threatening a shift to β (resulting in $\alpha\beta$) if he does. If North moves first, East can only lose by moving to β, and similarly if North can commit himself to α before East can make his threat;

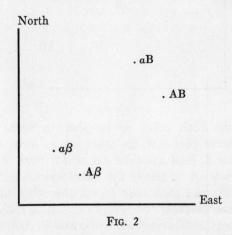

FIG. 2

but if East can effectively threaten the mutually undesirable $\alpha\beta$, he leaves North only a choice of $\alpha\beta$ or AB and North chooses the latter. Note that it is not sufficient for East to commit his *choice* in advance, as it was in Figure 1; he must commit himself to a *conditional* choice, B or β depending on whether North chooses A or α. If East committed his choice he would obtain only the advantage of "first move"; and in the present game, if moves were in turn, North would win at αB regardless of who moved first. (East would choose B rather than β, to leave North a choice of αB or AB rather than of $\alpha\beta$ or $A\beta$; and North would take αB. North, with first move, would choose α rather than A, leaving East $\alpha\beta$ or αB rather than $A\beta$ or AB; East would take αB.)

Figure 3 illustrates the promise. Whoever goes first, or even if moves are simultaneous, αB is a "minimax"; either can achieve it by himself, and neither can threaten the other with anything worse. Both would, however, prefer $A\beta$ to αB; but to reach $A\beta$

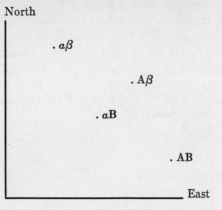

Fig. 3

they must trust each other or be able to make enforceable promises. Whoever goes first, the other has an incentive to cheat; if North chooses A, East can take AB, and if East chooses β first, North can choose $\alpha\beta$. If moves are simultaneous each has an incentive to cheat, and each may expect the other to cheat; and either deliberate cheating, or self-protection against the other's incentive to cheat, indicates choices of α and B. At least one party

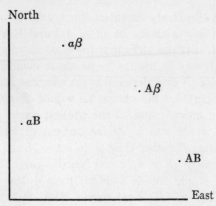

Fig. 4

must be able to commit himself to abstention; then the other can move first. If both must move simultaneously, both must be able to make enforceable promises.

Figure 4 is the same as Fig. 3 except that aB has been moved leftward. Here, in the absence of communication, North wins at $a\beta$ regardless of whether he or East moves first or moves are simultaneous. If, however, East can communicate a *conditional* commitment, he can force North to choose A and an outcome of $A\beta$. But this commitment is something more than either a promise or a threat; it is both a promise and a threat. He must threaten aB if North chooses a; and he must promise "not AB" if North chooses A. The threat alone will not induce North to avoid a; aB is better than AB for North, and AB is what he gets with A if East is free to choose B. East must commit himself to do, for either a or A, the opposite of what he would do if he were not committed: abstention from AB or immolation at aB.

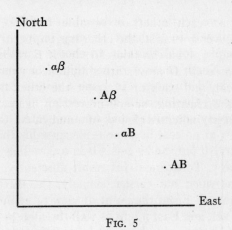

FIG. 5

Finally, Figs. 5 and 6 show two games that separately contain nothing of interest but together make possible an extortionate threat. Figure 5 has a minimax solution at aB; either can achieve aB, neither can enforce anything better, no collaboration is possible, no threat can be made. Figure 6, though contrasting with Fig. 5 in the identity of interest between the two parties, is similarly devoid of any need for collaboration or communication or any possible threat to exploit. With or without communication, with or without an order of moves, the outcome is at AB.

But suppose the two games are simultaneously up for decision,

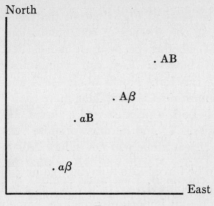

FIG. 6

and the same two parties are involved in both. If either party can commit himself to a threat he may improve his position. East, for example, could threaten to choose β rather than B in game 6, unless North chose A rather than a in game 5; alternatively, North could threaten a in game 6 unless East chose β in game 5. Assuming the intervals large enough in game 6, and the threat persuasively committed and communicated, the threatener gains in game 5 at no cost in game 6. Because his threat succeeds he does not carry it out; so he gets AB in 6 as well as his preferred choice in game 5. To express this result differently, game 6 supplies what was ruled out earlier, namely, the threat of an act "outside the game." From the point of view of game 5, game 6 is an extraneous act, and East might as well threaten to burn North's house down if he does not choose A in 5. But such purely extortionate threats are not always easy to make; they often require an occasion, an object, and a means of communication, and additionally often suffer from illegality, immorality, or resistance out of sheer stubbornness. The joining of two negotiations on the same agenda may thus succeed where a purely gratuitous threat would be impracticable.

If North cannot commit himself to a threat, and consequently desires only to prevent a threat by East, it is in his interest that communication be impossible; or if communication occurs, it is in his interest that the two games not be placed on the same agenda; or if he cannot prevent their being discussed together by East, it is in his interest to turn each game over to a different agent whose compensation depends only on the outcome of his own game. If North can force game 6 to be played first, and is unable to commit himself in response to a threat, the threat is obviated. If he can commit his choice in game 5 before the threat is made, he is safe. But if he can commit himself in game 5, and game 6 is to be played first, East could threaten to choose β in game 6 unless North assumed a prior commitment to A in game 5; in this case North's ability to commit himself is a disadvantage, since it permits him to be forced into "playing" game 5 ahead of 6.

Incidentally, dropping AB vertically in Fig. 2 to below the level $\alpha\beta$ would illustrate an important principle, namely, that moving one point in a manner "unfavorable" to North may actually improve the outcome for him. The threat that kept him from winning in Fig. 2 depends on the comparative attractiveness of AB over $\alpha\beta$ for North; if AB is made worse for him than $\alpha\beta$ he becomes immune to the threat, which then is not made, and he wins at $\alpha\beta$. This is an abstract example of the principle that, in bargaining, weakness may be strength.

3

BARGAINING, COMMUNICATION, AND LIMITED WAR

Limited war requires limits; so do strategic maneuvers if they are to be stabilized short of war. But limits require agreement or at least some kind of mutual recognition and acquiescence. And agreement on limits is difficult to reach, not only because of the uncertainties and the acute divergence of interests but because negotiation is severely inhibited both during war and before it begins and because communication becomes difficult between adversaries in time of war. Furthermore, it may seem to the advantage of one side to avoid agreement on limits, in order to enhance the other's fear of war; or one side or both may fear that even a show of willingness to negotiate will be interpreted as excessive eagerness.

The study of tacit bargaining — bargaining in which communication is incomplete or impossible — assumes importance, therefore, in connection with limited war, or, for that matter, with limited competition, jurisdictional maneuvers, jockeying in a traffic jam, or getting along with a neighbor that one does not speak to. The problem is to develop a modus vivendi when one or both parties either cannot or will not negotiate explicitly or when neither would trust the other with respect to any agreement explicitly reached. The present chapter will examine some of the concepts and principles that seem to underlie tacit bargaining and will attempt to draw a few illustrative conclusions about the problem of limited war or analogous situations. It will also suggest that these same principles may often provide a powerful clue to understanding even the logically dissimilar case of explicit bargaining with full communication and enforcement.

The most interesting situations and the most important are those in which there is a conflict of interest between the parties involved. But it is instructive to begin with the special simplified case in which two or more parties have identical interests and face the problem not of reconciling interests but only of coordinating their actions for their mutual benefit, when communication is impossible. This special case brings out clearly the principle that will then serve to solve the problem of tacit "bargaining" over conflicting preferences.

TACIT COORDINATION (COMMON INTERESTS)

When a man loses his wife in a department store without any prior understanding on where to meet if they get separated, the chances are good that they will find each other. It is likely that each will think of some obvious place to meet, so obvious that each will be sure that the other is sure that it is "obvious" to both of them. One does not simply predict where the other will go, since the other will go where he predicts the first to go, which is wherever the first predicts the second to predict the first to go, and so ad infinitum. Not "What would I do if I were she?" but "What would I do if I were she wondering what she would do if she were I wondering what I would do if I were she . . . ?" What is necessary is to coordinate predictions, to read the same message in the common situation, to identify the one course of action that their expectations of each other can converge on. They must "mutually recognize" some unique signal that coordinates their expectations of each other. We cannot be sure they will meet, nor would all couples read the same signal; but the chances are certainly a great deal better than if they pursued a random course of search.

The reader may try the problem himself with the adjoining map (Fig. 7). Two people parachute unexpectedly into the area shown, each with a map and knowing the other has one, but neither knowing where the other has dropped nor able to communicate directly. They must get together quickly to be rescued. Can they study their maps and "coordinate" their behavior? Does the map suggest some particular meeting place so unambiguously

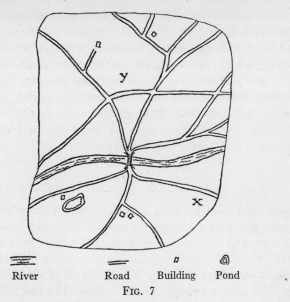

| River | Road | Building | Pond |

Fig. 7

that each will be confident that the other reads the same suggestion with confidence?

The writer has tried this and other analogous problems on an unscientific sample of respondents; and the conclusion is that people often can coordinate. The following abstract puzzles are typical of those that can be "solved" by a substantial proportion of those who try. The solutions are, of course, arbitrary to this extent: any solution is "correct" if enough people think so. The reader may wish to confirm his ability to concert in the following problems with those whose scores are given in a footnote.[1]

[1] In the writer's sample, 36 persons concerted on "heads" in problem 1, and only 6 chose "tails." In problem 2, the first three numbers were given 37 votes out of a total of 41; the number 7 led 100 by a slight margin, with 13 in third place. The upper left corner in problem 3 received 24 votes out of a total of 41, and all but 3 of the remainder were distributed in the same diagonal line. Problem 4, which may reflect the location of the sample in New Haven, Connecticut, showed an absolute majority managing to get together at Grand Central Station (information booth), and virtually all of them succeeded in meeting at 12 noon. Problem 6 showed a variety of answers, but two-fifths of all persons succeeded in concerting on the number 1; and in problem 7, out of 41 people, 12 got together on $1,000,000, and only 3 entries consisted of numbers that were not a power of 10; of those 3, 2 were $64 and, in the

1. Name "heads" or "tails." If you and your partner name the same, you both win a prize.

2. Circle one of the numbers listed in the line below. You win if you all succeed in circling the same number.

    7   100   13   261   99   555

3. Put a check mark in one of the sixteen squares. You win if you all succeed in checking the same square.

☐ ☐ ☐ ☐

☐ ☐ ☐ ☐

☐ ☐ ☐ ☐

☐ ☐ ☐ ☐

4. You are to meet somebody in New York City. You have not been instructed where to meet; you have no prior understanding with the person on where to meet; and you cannot communicate with each other. You are simply told that you will have to guess where to meet and that he is being told the same thing and that you will just have to try to make your guesses coincide.

5. You were told the date but not the hour of the meeting in No. 4; the two of you must guess the exact minute of the day for meeting. At what time will you appear at the meeting place that you elected in No. 4?

6. Write some positive number. If you all write the same number, you win.

7. Name an amount of money. If you all name the same amount, you can have as much as you named.

8. You are to divide $100 into two piles, labeled A and B.

more up-to-date version, $64,000! Problem 8 caused no difficulty to 36 out of 41, who split the total fifty-fifty. Problem 9 secured a majority of 20 out of 22 for Robinson. An alternative formulation of it, in which Jones and Robinson were tied on the first ballot at 28 votes each, was intended by the author to demonstrate the difficulty of concerting in case of tie; but the respondents surmounted the difficulty and gave Jones 16 out of 18 votes (apparently on the basis of Jones's earlier position on the list), proving the main point but overwhelming the subsidiary point in the process. In the map most nearly like the one reproduced here (Fig. 1), 7 out of 8 respondents managed to meet at the bridge.

Your partner is to divide another $100 into two piles labeled A and B. If you allot the same amounts to A and B, respectively, that your partner does, each of you gets $100; if your amounts differ from his, neither of you gets anything.

9. On the first ballot, candidates polled as follows:

Smith	19	Robinson	29
Jones	28	White	9
Brown	15		

The second ballot is about to be taken. You have no interest in the outcome, except that you will be rewarded if someone gets a majority on the second ballot and you vote for the one who does. Similarly, all voters are interested only in voting with the majority, and everybody knows that this is everybody's interest. For whom do you vote on the second ballot?

These problems are artificial, but they illustrate the point. People *can* often concert their intentions or expectations with others if each knows that the other is trying to do the same. Most situations — perhaps every situation for people who are practiced at this kind of game — provide some clue for coordinating behavior, some focal point for each person's expectation of what the other expects him to expect to be expected to do. Finding the key, or rather finding *a* key — any key that is mutually recognized as the key becomes *the* key — may depend on imagination more than on logic; it may depend on analogy, precedent, accidental arrangement, symmetry, aesthetic or geometric configuration, casuistic reasoning, and who the parties are and what they know about each other. Whimsy may send the man and his wife to the "lost and found"; or logic may lead each to reflect and to expect the other to reflect on where they would have agreed to meet if they had had a prior agreement to cover the contingency. It is not being asserted that they will always find an obvious answer to the question; but the chances of their doing so are ever so much greater than the bare logic of abstract random probabilities would ever suggest.

A prime characteristic of most of these "solutions" to the problems, that is, of the clues or coordinators or focal points, is some kind of prominence or conspicuousness. But it is a prominence that depends on time and place and who the people are. Ordinary folk lost on a plane circular area may naturally go to the center to meet each other; but only one versed in mathematics would "naturally" expect to meet his partner at the center of gravity of an irregularly shaped area. Equally essential is some kind of uniqueness; the man and his wife cannot meet at the "lost and found" if the store has several. The writer's experiments with alternative maps indicated clearly that a map with many houses and a single crossroads sends people to the crossroads, while one with many crossroads and a single house sends most of them to the house. Partly this may reflect only that uniqueness conveys prominence; but it may be more important that uniqueness avoids ambiguousness. Houses may be intrinsically more prominent than anything else on the map; but if there are three of them, none more prominent than the others, there is but one chance in three of meeting at a house, and the recognition of this fact may lead to the rejection of houses as the "clue." [2]

But in the final analysis we are dealing with imagination as much as with logic; and the logic itself is of a fairly casuistic kind. Poets may do better than logicians at this game, which is perhaps more like "puns and anagrams" than like chess. Logic helps — the large plurality accorded to the number 1 in problem 6 seems to rest on logic — but usually not until imagination has selected some clue to work on from among the concrete details of the situation.

TACIT BARGAINING (DIVERGENT INTERESTS)

A conflict of interest enters our problem if the parachutists dislike walking. With communication, which is not allowed in our problem, they would have argued or bargained over where to meet, each favoring a spot close to himself or a resting place particularly to his liking. In the absence of communication, their overriding interest is to concert ideas; and if a particular spot

[2] That this would be "correct" reasoning, incidentally, is suggested by one of the author's map experiments. On a map with a single house and many crossroads, the eleven people who chose the house all met, while the four who chose crossroads all chose different crossroads and did not even meet one another.

commands attention as the "obvious" place to meet, the winner of the bargain is simply the one who happens to be closer to it. Even if the one who is farthest from the focal point knows that he is, he cannot withhold his acquiescence and argue for a fairer division of the walking; the "proposal" for the bargain that is provided by the map itself — if, in fact, it provides one — is the only extant offer; and without communication, there is no counterproposal that can be made. The conflict gets reconciled — or perhaps we should say ignored — as a by-product of the dominant need for coordination.

"Win" and "lose" may not be quite accurate, since both may lose by comparison with what they could have agreed on through communication. If the two are actually close together and far from the lone house on the map, they might have eliminated the long walk to the house if they could have identified their locations and concerted explicitly on a place to meet between them. Or it may be that one "wins" while the other loses more than the first wins: if both are on the same side of the house and walk to it, they walk together a greater distance than they needed to, but the closer one may still have come off better than if he had had to argue it out with the other.

This last case illustrates that it may be to the advantage of one to be unable to communicate. There is room here for a motive to destroy communication or to refuse to collaborate in advance on a method of meeting if one is aware of his advantage and confident of the "solution" he foresees. In one variant of the writer's test, A knew where B was, but B had no idea where A was (and each knew how much the other knew). Most of the recipients of the B-type questionnaire smugly sat tight, enjoying their ignorance, while virtually all the A-questionnaire respondents grimly acknowledged the inevitable and walked all the way to B. Better still may be to have the power to send but not to receive messages: if one can announce his position and state that his transmitter works but not his receiver, saying that he will wait where he is until the other arrives, the latter has no choice. He can make no effective counteroffer, since no counteroffer could be heard.[3]

[3] This is an instance of the general paradox, illustrated at length in Chap-

The writer has tried a sample of conflicting-interest games on a number of people, including games that are biased in favor of one party or the other; and on the whole, the outcome suggests the same conclusion that was reached in the purely cooperative games. All these games require coordination; they also, however, provide several alternative choices over which the two parties' interests differ. Yet, among all the available options, some particular one usually seems to be the focal point for coordinated choice, and the party to whom it is a relatively unfavorable choice quite often takes it simply because he knows that the other will expect him to. The choices that cannot coordinate expectations are not really "available" without communication. The odd characteristic of all these games is that neither rival can gain by outsmarting the other. Each loses unless he does exactly what the other expects him to do. Each party is the prisoner or the beneficiary of their mutual expectations; no one can disavow his own expectation of what the other will expect him to expect to be expected to do. The need for agreement overrules the potential disagreement, and each much concert with the other or lose altogether. Some of these games are arrived at by slightly changing the problems given earlier, as we did for the map problem by supposing that walking is onerous.

1. A and B are to choose "heads" or "tails" without communicating. If both choose "heads," A gets $3 and B gets $2; if both choose "tails," A gets $2 and B gets $3. If they choose differently, neither gets anything. You are A (or B); which do you choose? (Note that if both choose at random, there is only a 50-50 chance of successful coincidence and an expected value of $1.25 apiece — less than either $3 or $2.)

2. You and your two partners (or rivals) each have one of the letters A, B, and C. Each of you is to write these three letters, A, B, and C, in any order. If the order is the same on all three of your lists, you get prizes totaling $6, of which $3 goes to the one whose letter is first on all three lists, $2 to the one whose letter is second, and $1 to the person whose letter is third. If the letters are not in identical order on all three lists, none of

---

ter 2, that what is impotence by ordinary standards may, in bargaining, be a source of "strength."

you gets anything. Your letter is A (or B, or C); write here the three letters in the order you choose:

———, ———, ———.

3. You and your partner (rival) are each given a piece of paper, one blank and the other with an "X" written on it. The one who gets the "X" has the choice of leaving it alone or erasing it; the one who gets the blank sheet has the choice of leaving it blank or writing an "X" on it. If, when you have made your choices without communicating, there is an "X" on only one of the sheets, the holder of the "X" gets $3 and the holder of the blank sheet gets $2. If both sheets have "X's" or both sheets are blank, neither gets anything. Your sheet of paper has the original "X" on it; do you leave it alone or erase it? (*Alternate:* your sheet of paper is the blank one; do you leave it blank or write an "X"?)

4. You and your partner (rival) are to be given $100 if you can agree on how to divide it without communicating. Each of you is to write the amount of his claim on a sheet of paper; and if the two claims add to no more than $100, each gets exactly what he claimed. If the two claims exceed $100, neither of you gets anything. How much do you claim? $———.

5. You and your partner are each to pick one of the five letters, K, G, W, L, or R. If you pick the same letter, you get prizes; if you pick different letters, you get nothing. The prizes you get depend on the letter you both pick; but the prizes are not the same for each of you, and the letter that would yield you the highest prize may or may not be his most profitable letter. For you the prizes would be as follows:

K	$4	L	$2
G	$3	R	$5
W	$1		

You have no idea what his schedule of prizes looks like. You begin by proposing to him the letter R, that being your best letter. Before he can reply, the master-of-ceremonies intervenes to say that you were not supposed to be allowed to communicate and that any further communication will disqualify you both. You must simply write down one of the letters, hoping that the other chooses the same letter. Which letter do you choose? (Alternate formulation for the second half of the sample shows schedule of K–$3, G–$1, W–$4, L–$5, R–$2, and has the "other" party make the initial proposal of the letter R before communication is cut off.)

6. Two opposing forces are at the points marked X and Y in a map similar to the one in Fig. 1. The commander of each force wishes to occupy as much of the area as he can and knows the other does too. But each commander wishes to avoid an armed clash and knows the other does too. Each must send forth his troops with orders to take up a designated line and to fight if opposed. Once the troops are dispatched, the outcome depends only on the lines that the two commanders have ordered their troops to occupy. If the lines overlap, the troops will be assumed to meet and fight, to the disadvantage of both sides. If the troops take up positions that leave any appreciable space unoccupied between them, the situation will be assumed "unstable" and a clash inevitable. Only if the troops are ordered to occupy identical lines or lines that leave virtually no unoccupied space between them will a clash be avoided. In that case, each side obtains successfully the area it occupies, the advantage going to the side that has the most valuable area in terms of land and facilities. You command the forces located at the point marked X (Y). Draw on the map the line that you send your troops to occupy.

7. A and B have incomes of $100 and $150 per year, respectively. They are notified of each other's income and told that they must begin paying taxes totaling $25 per year. If they can reach agreement on shares of this total, they may share the annual tax bill in whatever manner they agree on. But they must reach agreement without communication; each is to write down the share he proposes to pay, and if the shares total $25 or more, each will pay exactly what he proposed. If the proposed shares fail to add up to $25, however, each will individually be required to pay the full $25, and the tax collectors will keep the surplus. You are A (B); how much do you propose to pay? $_____.

8. A loses some money, and B finds it. Under the house rules, A cannot have his money back until he agrees with the finder on a suitable reward, and B cannot keep any except what A agrees to. If no agreement is reached, the money goes to the house. The

amount is $16, and A offers $2 as a reward. B refuses, demanding half the money for himself. An argument ensues, and the house intervenes, insisting that each write his claim, once and for all, without further communication. If the claims are consistent with the $16 total, each will receive exactly what he claims; but if together they claim more than $16, the funds will be confiscated by the house. As they sit pondering what claims to write, a well-known and respected mediator enters and offers to help. He cannot, he says, participate in any bargaining, but he can make a "fair" proposal. He approaches A and says, "I think a reasonable division under the circumstances would be a 2-1 split, the original owner getting two-thirds and the finder one-third, perhaps rounded off to $11 and $5, respectively. I shall make the same suggestion to him." Without waiting for any response, he approaches the finder, makes the same suggestion, and says that he made the same suggestion to the original owner. Again without waiting for any response, he departs. You are A (B); what claim do you write?

The outcomes in the writer's informal sample are given in the footnote.[4] In those problems where there is some asymmetry between "you" and "him," that is, between A and B, the A formulations were matched with the B formulations in deriving

[4] In the first problem, 16 out of 22 A's and 15 out of 22 B's chose heads. Given what the A's did, heads was the best answer for B; given what the B's did, heads was the best answer for A. Together they did substantially better than at random; and, of course, if each had tried to win $3, they would all have scored a perfect zero. Problem 2, however, which is logically similar to 1 but with a more compelling structure, showed 9 out of 12 A's, 10 out of 12 B's, and 14 out of 16 C's, successfully co-ordinating on ABC. (Of the remaining 7, incidentally, 5 discriminated against themselves in departing from alphabetical order, all to no avail.) Problem 3, which is structurally analogous to 1, showed 18 out of 22 A's concerting successfully with 14 out of 19 B's, giving A the $3 prize. In problem 4, 36 out of 40 chose $50. (Two of the remainder were $49 and $49.99.) In problem 5 the letter R won 5 out of 8 votes from those who had proposed it, and 8 out of 9 votes from those who were on the other side. In problem 6, 14 of 22 X's and 14 of 23 Y's drew their boundaries exactly along the river. The "correctness" of this solution is emphatically shown by the fact that the other 15, who eschewed the river, produced 14 different lines. Of 8 × 7 possible pairs among them, there were 55 failures and 1 success. Problem 7 showed 5 out of 6 of those with incomes of $150 and 7 out of 10 of those with incomes of $100 concerting on a 15–10 division of the tax. In problem 8 both those who lost money and those who found it, 8 and 7 persons respectively, unanimously concerted on the mediator's suggestion of an even $5 reward.

the "outcome." The general conclusion, as given in more detail in the footnote, is that the participants can "solve" their problem in a substantial proportion of the cases; they certainly do conspicuously better than any chance methods would have permitted, and even the disadvantaged party in the biased games permits himself to be disciplined by the message that the game provides for their coordination.

The "clues" in these games are diverse. Heads apparently beat tails through some kind of conventional priority, similar to the convention that dictates A, B, C, though not nearly so strong. The original X beats the blank sheet, apparently because the "status quo" is more obvious than change. The letter R wins because there is nothing to contradict the first offer. Roads might seem, in principle, as plausible as rivers, especially since their variety permits a less arbitrary choice. But, precisely because of their variety, the map cannot say *which* road; so roads must be discarded in favor of the unique and unambiguous river. (Perhaps in a symmetrical map of uniform terrain, the outcome would be more akin to the 50-50 split in the $100 example — a diagonal division in half, perhaps — but the irregularity of the map rather precludes a geometrical solution.)

The tax problem illustrates a strong power of suggestion in the income figures. The abstract logic of this problem is identical with that of the $100 division; in fact, it could be reworded as follows: each party pays $25 in taxes, and a refund of $25 is available to be divided among the two parties if they can agree on how to divide it. This formulation is logically equivalent to the one in problem 7, and, as such, it differs from problem 4 only in the amount of $25 instead of $100. Yet the inclusion of income figures, just by *suggesting* their relevance and making them prominent in the problem, shifts the focal point substantially to a 10-15 split rather than 12.5-12.5. And why, if incomes are relevant, is a perfectly *proportional* tax so obvious, when perhaps there are grounds for graduated rates? The answer must be that no *particular* graduation of rates is so obvious as to go without saying; and if speech is impossible, by default the uniquely simple and recognizable principle of proportionality has to be adopted. First the income figures take the initial plausibility away from a 50-50 split; then the simplicity of proportionality makes 10-15 the only one that could possibly be considered capable of tacit recognition. The same principle is displayed by an experiment in which question 7 was deliberately cluttered up with *additional* data — on family size, spending habits, and so on. Here the unique attraction of the income-proportionate split apparently became so diluted that the preponderant reply from both the high-income and the low-income respondents was a simple 50-50 division of the tax. The refined signal for the income proportionate split was drowned out by "noise," and the cruder signal for equality was all that came through.

Finally, problem 8 is again logically the same as problem 4, the amount being $16 available for two people if they can write claims that do not exceed the amount. But the institutional arrangement is discriminatory; finder and loser do not have a compelling equality in any moralistic or legalistic sense, so the 50-50 split seems not quite obvious. The suggestion of the mediator provides the only other signal that is visible; its potency as a coordinator is seen even in the rounding to $11 and $5, which was universally accepted.

In each of these situations the outcome is determined by something that is fairly arbitrary. It is not a particularly "fair" outcome, from either an observer's point of view or the points of view of the participants. Even the 50-50 split is arbitrary in its reliance on a kind of recognizable mathematical purity; and if it is "fair," it is so only because we have no concrete data by which to judge its unfairness, such as the source of the funds, the relative need of the rival claimants, or any potential basis for moral or legal claims. Splitting the difference in an argument over kidnap ransom is not particularly "fair," but it has the mathematical qualities of problem 4.

If we ask what determines the outcome in these cases, the answer again is in the coordination problem. Each of these problems requires coordination for a common gain, even though there is rivalry among alternative lines of common action. But, among the various choices, there is usually one or only a few than can serve as coordinator. Take the case of the first offer in problem 5. The strongest argument in favor of R is the rhetorical question,

"If not R, what then?" There is no answer so obvious as to give more than a random chance of concerting, even if both parties wanted to eschew the letter R after the first offer was made. To illustrate the force of this point, suppose that the master-of-ceremonies in that problem considered the first offer already to have spoiled the game and thought he might confuse the players by announcing the reversal of their prize schedules. A will get whatever prize B would have gotten, and B will get the prizes shown in A's schedule in problem 5. Does the original offerer of R have any reason to change his choice? Or suppose that the master-of-ceremonies announced that the prizes would be the same, no matter what letter were chosen, so long as they both picked the same letter. They will still rally to R as the only indicated means of coordinating choices. If we revert to the beginning of this game and suppose that the original proposal of R never got made, we might imagine a sign on the wall saying, "In case of doubt always choose R; this sign is visible to all players and constitutes a means of coordinating choices." Here we are back at the man and his wife in the department store, whose problems are over when they see a conspicuous sign that says, "The management suggests that all persons who become separated meet each other at the information booth in the center of the ground floor." Beggars cannot be choosers about the source of their signal, or about its attractiveness compared with others that they can only wish were as conspicuous.

The irony would be complete if, in game 5, your rival knew your prize schedule and you did not know his (as was the case in a variant of question 5 used in some questionnaires). Since you have no basis for guessing his preference and could not even do him a favor or make a "fair" compromise if you wished to, the only basis for concerting is to see what message you can both read in your schedule. Your own preferred letter seems the indicated choice; it is hard to see why to pick any other or which other to pick, since you have no basis for knowing what other letter is better for him than R itself. His knowledge of your preference, combined with your ignorance of his and the lack of any alternative basis for coordination, puts on him the responsibility of simply choosing in your favor. (This, in fact, was the preponderant result among the small sample tested.) It is the same situation as when only one parachutist knew where the other was.[5]

EXPLICIT BARGAINING

The concept of "coordination" that has been developed here for tacit bargaining does not seem directly applicable to explicit bargaining. There is no apparent need for intuitive rapport when speech can be used; and the adventitious clues that coordinated thoughts and influenced the outcome in the tacit case revert to the status of incidental details.

Yet there is abundant evidence that some such influence is powerfully present even in explicit bargaining. In bargains that involve numerical magnitudes, for example, there seems to be a strong magnetism in mathematical simplicity. A trivial illustration is the tendency for the outcomes to be expressed in "round numbers"; the salesman who works out the arithmetic for his "rock-bottom" price on the automobile at $2,507.63 is fairly pleading to be relieved of $7.63. The frequency with which final agreement is precipitated by an offer to "split the difference" illustrates the same point, and the difference that is split is by no means always trivial. More impressive, perhaps, is the remarkable frequency with which long negotiations over complicated quantitative formulas or *ad hoc* shares in some costs or benefits converge ultimately on something as crudely simple as equal shares, shares proportionate to some common magnitude (gross national product, population, foreign-exchange deficit, and so forth), or the shares agreed on in some previous but logically irrelevant negotiation.[6]

Precedent seems to exercise an influence that greatly exceeds its logical importance or legal force. A strike settlement or an international debt settlement often sets a "pattern" that is fol-

[5] And it is another example of the power that resides in "weakness," which was commented on in an earlier footnote.

[6] From a great variety of formulas proposed for the contributions to UNRRA, the winner that emerged was a straight 1 per cent of gross national product — the simplest conceivable formula and the roundest conceivable number. This formula was, to be sure, the preferred position of the United States during the discussion; but that fact perhaps adds as much to the example as it detracts from it.

lowed almost by default in subsequent negotiations. Sometimes, to be sure, there is a reason for a measure of uniformity, and sometimes there is enough similarity in the circumstances to explain similar outcomes; but more often it seems that there is simply no heart left in the bargaining when it takes place under the shadow of some dramatic and conspicuous precedent.[7] In similar fashion, mediators often display a power to precipitate agreement and a power to determine the terms of agreement; their proposals often seem to be accepted less by reason of their inherent fairness or reasonableness than by a kind of resignation of both participants. "Fact-finding" reports may also tend to draw expectations to a focus, by providing a suggestion to fill the vacuum of indeterminacy that otherwise exists: it is not the facts themselves, but the creation of a specific suggestion, that seems to exercise the influence.

There is, in a similar vein, a strong attraction to the *status quo ante* as well as to natural boundaries. Even parallels of latitude have recently exhibited their longevity as focal points for agreement. Certainly there are reasons of convenience in using rivers as the agreed stopping place for troops or using old boundaries, whatever their current relevance; but often these features of the landscape seem less important for their practical convenience than for their power to crystallize agreement.

These observations would be trivial if they meant only that bargaining results were *expressed* in simple and qualitative terms or that minor accommodations were made to round off the last few cents or miles or people. But it often looks as though the ultimate focus for agreement did not just reflect the balance of bargaining powers but provided bargaining power to one side or the other. It often seems that a cynic could have predicted the outcome on the basis of some "obvious" focus for agreement, some strong suggestion contained in the situation itself, without much regard to the merits of the case, the arguments to be made, or the pressures to be applied during the bargaining. The "obvious" place to compromise frequently seems to win by some

---

[7] This and the preceding paragraph are illustrated by the speed with which a number of Middle Eastern oil-royalty arrangements converged on the 50-50 formula a few years after World War II.

kind of default, as though there is simply no rationale for settling anywhere else. Or, if the "natural" outcome is taken to
reflect the relative skills of the parties to the bargain, it may be
important to identify that skill as the ability to set the stage
in such a way as to give prominence to some particular outcome
that would be favorable. The outcome may not be so much conspicuously fair or conspicuously in balance with estimated bargaining powers as just plain "conspicuous."

This conclusion may seem to reduce the scope for bargaining
skill, if the outcome is already determined by the configuration
of the problem itself and where the focal point lies. But perhaps
what it does is shift the locus where skill is effective. The "obvious" outcome depends greatly on how the problem is formulated,
on what analogies or precedents the definition of the bargaining
issue calls to mind, on the kinds of data that may be available to
bear on the question in dispute. When the committee begins to
argue over how to divide the costs, it is already constrained by
whether the terms of reference refer to the "dues" to be shared or
the "taxes" to be paid, by whether a servicing committee is preparing national-income figures or balance-of-payments figures for
their use, by whether the personnel of the committee brings certain
precedents into prominence by having participated personally in
earlier negotiations, by whether the inclusion of two separate
issues on the same agenda will give special prominence and relevance to those particular features that they have in common.
Much of the skill has already been applied when the formal negotiations begin.[8]

If all this is correct, as it seems frequently to the author to be,
our analysis of tacit bargaining may help to provide an understanding of the influence at work; and perhaps the logic of tacit
bargaining even provides a basis for believing it to be correct.
The fundamental problem in tacit bargaining is that of *coordination*; we should inquire, then, what has to be coordinated in ex-

[8] Perhaps another role for skill is contained in this general approach. If one is
unsuccessful in getting the problem so formulated that the "obvious" outcome
is near his own preferred position, he can proceed to confuse the issue. Find
multiple definitions for all the terms and add "noise" to drown out the strong
signal contained in the original formulation. The technique may not succeed,
but in the variant of our income-tax problem mentioned above it certainly did.

plicit bargaining. The answer may be that explicit bargaining requires, for an ultimate agreement, some coordination of the participants' expectations. The proposition might be as follows.

Most bargaining situations ultimately involve some range of possible outcomes within which each party would rather make a concession than fail to reach agreement at all. In such a situation any potential outcome is one from which at least one of the parties, and probably both, would have been willing to retreat for the sake of agreement, and very often the other party knows it. Any potential outcome is therefore one that either party could have improved by insisting; yet he may have no basis for insisting, since the other knows or suspects that he would rather concede than do without agreement. Each party's strategy is guided mainly by what he expects the other to accept or insist on; yet each knows that the other is guided by reciprocal thoughts. The final outcome must be a point from which neither expects the other to retreat; yet the main ingredient of this expectation is what one thinks the other expects the first to expect, and so on. Somehow, out of this fluid and indeterminate situation that seemingly provides no logical reason for anybody to expect anything except what he expects to be expected to expect, a decision is reached. These infinitely reflexive expectations must somehow converge on a single point, at which each expects the other not to expect to be expected to retreat.

If we then ask what it is that can bring their expectations into convergence and bring the negotiation to a close, we might propose that it is the intrinsic magnetism of particular outcomes, especially those that enjoy prominence, uniqueness, simplicity, precedent, or some rationale that makes them qualitatively differentiable from the continuum of possible alternatives. We could argue that expectations tend not to converge on outcomes that differ only by degree from alternative outcomes but that people have to dig in their heels at a groove in order to make any show of determination. One has to have a reason for standing firmly on a position; and along the continuum of qualitatively undifferentiable positions one finds no rationale. The rationale may not be strong at the arbitrary "focal point," but at least it can defend itself with the argument "If not here, where?"

There is perhaps a little more to this need for a mutually identifiable resting place. If one is about to make a concession, he needs to control his adversary's expectations; he needs a recognizable limit to his own retreat. If one is to make a finite concession that is not to be interpreted as capitulation, he needs an obvious place to stop. A mediator's suggestion may provide it; or any other element that qualitatively distinguishes the new position from surrounding positions. If one has been demanding 60 per cent and recedes to 50 per cent, he can get his heels in; if he recedes to 49 per cent, the other will assume that he has hit the skids and will keep sliding.

If some troops have retreated to the river in our map, they will expect to be expected to make a stand. This is the one spot to which they can retreat without necessarily being expected to retreat further, while, if they yield any further, there is no place left where they can be expected to make a determined stand. Similarly, the advancing party can expect to force the other to retreat to the river without having his advance interpreted as an insatiable demand for unlimited retreat. There is stability at the river — and perhaps nowhere else.

This proposition may seem intuitively plausible; it does to the writer, and in any event some kind of explanation is needed for the tendency to settle at focal points. But the proposition would remain vague and somewhat mystical if it were not for the somewhat more tangible logic of tacit bargaining. The latter provides not only an analogy but the demonstration that the necessary psychic phenomenon — tacit coordination of expectations — is a real possibility and in some contexts a remarkably reliable one. The "coordination" of expectations is analogous to the "coordination" of behavior when communication is cut off; and, in fact, they both involve nothing more nor less than intuitively perceived mutual expectations. Thus the empirically verifiable results of some of the tacit-bargaining games, as well as the more logical role of coordinated expectations in that case, prove that expectations can be coordinated and that some of the objective details of the situation can exercise a controlling influence when the coordination of expectations is essential. *Something* is perceived by both parties when communication is absent;

it must still be perceptible, though undoubtedly of lesser force, when communication is possible. The possibility of communication does not make 50-50 less symmetrical or the river less unique or A B C a less natural order for those letters.

If all we had to reason from were the logic of tacit bargaining, it would be only a guess and perhaps a wild one that the same kind of psychic attraction worked in explicit bargaining; and if all we had to generalize from were the observation of peculiarly "plausible" outcomes in actual bargains, we might be unwilling to admit the force of adventitious details. But the two lines of evidence so strongly reinforce each other that the analogy between tacit and explicit bargaining seems a potent one.

To illustrate with the problem of agreeing explicitly on how to divided $100: 50-50 seems a plausible division, but it may seem so for too many reasons. It may seem "fair"; it may seem to balance bargaining powers; or it may, as suggested in this paper, simply have the power to communicate its own inevitability to the two parties in such fashion that each appreciates that they both appreciate it. What our analysis of tacit bargaining provides is evidence for the latter view. The evidence is simply that *if* they had to divide the $100 without communicating, they could concert on 50-50. Instead of relying on intuition, then, we can point to the fact that in a slightly different context — the tacit-bargaining context — our argument has an objectively demonstrable interpretation.

To illustrate again: the ability of the two commanders in one of our problems to recognize the stabilizing power of the river — or, rather, their inability not to recognize it — is substantiated by the evidence that if their survival depended on some agreement about where to stabilize their lines *and communication were not allowed,* they probably could perceive and appreciate the qualities of the river as a focus for their tacit agreement. So the tacit analogy at least demonstrates that the idea of "coordinating expectations" is meaningful rather than mystical.

Perhaps we could push the argument further still. Even in those cases in which the only distinguishing characteristic of a bargaining result is its evident "fairness," by standards that the participants are known to appreciate, we might argue that the moral force of fairness is greatly reinforced by the power of a "fair" result to focus attention, if it fills the vacuum of indeterminacy that would otherwise exist. Similarly, when the pressure of public opinion seems to force the participants to the obviously "fair" or "reasonable" solution, we may exaggerate the "pressure" or at least misunderstand the way it works on the participants unless we give credit to its power to coordinate the participants' expectations. It may, to put it differently, be the power of *suggestion,* working through the mechanism described in this paper, that makes public opinion or precedent or ethical standards so effective. Again, as evidence for this view, we need only to suppose that the participants had to reach ultimate agreement without communicating and visualize public opinion or some prominent ethical standard as providing a strong suggestion analogous to the suggestions contained in our earlier examples. The mediator in problem 7 is a close analogy. Finally, even if it is truly the force of moral responsibility or sensitivity to public opinion that constrains the participants, and not the "signal" they get, we must still look to the source of the public's own opinion; and there, the writer suggests, the need for a simple, qualitative rationale often reflects the mechanism discussed in this paper.

But, if this general line of reasoning is valid, any analysis of explicit bargaining must pay attention to what we might call the "communication" that is inherent in the bargaining situations, the signals that the participants read in the inanimate details of the case. And it means that tacit and explicit bargaining are not thoroughly separate concepts but that the various gradations from tacit bargaining up through types of incompleteness or faulty or limited communication to full communication all show some dependence on the need to coordinate expectations. Hence all show some degree of dependence of the participants themselves on their common inability to keep their eyes off certain outcomes.

This is not necessarily an argument for expecting explicit outcomes as a rule to lean toward exactly those that would have emerged if communication had been impossible; the focal points may certainly be different when speech is allowed, except in some of the artificial cases we have used in our illustrations. But what may be the *main* principle in tacit bargaining apparently may be at least *one* of the important principles in the analysis of explicit bargaining. And, since even much so-called "explicit" bargaining includes maneuver, indirect communication, jockeying for position, or speaking to be overheard, or is confused by a multitude of participants and divergent interests, the need for convergent expectations and the role of signals that have the power to coordinate expectations may be powerful.

Perhaps many kinds of social stability and the formation of interest groups reflect the same dependence on such coordinators as the terrain and the circumstances can provide: the band wagon at political conventions that often converts the slightest sign of plurality into an overwhelming majority; the power of constitutional legitimacy to command popular support in times of anarchy or political vacuum; the legendary power of an old gang leader to bring order into the underworld, simply because obedience depends on the expectation that others will be obedient in punishing disobedience. The often expressed idea of a "rallying point" in social action seems to reflect the same concept. In economics the phenomena of price leadership, various kinds of nonprice competition, and perhaps even price stability itself appear amenable to an analysis that stresses the importance of tacit communication and its dependence on qualitatively identifiable and fairly unambiguous signals that can be read in the situation itself. "Spontaneous" revolt may reflect similar principles: when leaders can easily be destroyed, people require some signal for their coordination, a signal so unmistakably comprehensible and so potent in its suggestion for action that everyone can be sure that everyone else reads the same signal with enough confidence to act on it, thus providing one another with the immunity that goes with action in large numbers. (There is even the possibility that such a signal might be provided from outside, even by an agent whose only claim to leadership was its capacity to signal the instructions required for concerted action.)

TACIT NEGOTIATION AND LIMITED WAR

What useful insight does this line of analysis provide into the practical problems of tacit bargaining that usually confront us, particularly the problems of strategic maneuver and limited war? It certainly suggests that it is *possible* to find limits to war — real war, jurisdictional war, or whatever — without overt negotiation. But it gives us no new strong sense of *probability*. War was limited in Korea, and gas was not used in World War II; on the possibility of limited war these two facts are more persuasive than all the suggestions contained in the foregoing discussion. If the analysis provides anything, then, it is not a judgment of the probability of successfully reaching tacit agreement but a better understanding of where to look for the terms of agreement.

If there are important conclusions to be drawn, they are probably these: (1) tacit agreements or agreements arrived at through partial or haphazard negotiation require terms that are qualitatively distinguishable from the alternatives and cannot simply be a matter of degree; (2) when agreement must be reached with incomplete communication, the participants must be ready to allow the situation itself to exercise substantial constraint over the outcome; specifically, a solution that discriminates against one party or the other or even involves "unnecessary" nuisance to both of them may be the only one on which their expectations can be coordinated.

Gas was not used in World War II. The agreement, though not without antecedents, was largely a tacit one. It is interesting to speculate on whether any alternative agreement concerning poison gas could have been arrived at without formal communication (or even, for that matter, with communication). "Some gas" raises complicated questions of how much, where, under what circumstances: "no gas" is simple and unambiguous. Gas only on military personnel; gas used only by defending forces; gas only when carried by vehicle or projectile; no gas without warning — a variety of limits is conceivable; some may make sense, and many might have been more impartial to the outcome of the war. But there is a simplicity to "no gas" that makes it almost uniquely a focus for agreement when each side can only conjecture at what rules the other side would propose and when failure at coordination on the first try may spoil the chances for acquiescence in any limits at all.

The physical configuration of Korea must have helped in defining the limits to war and in making geographical limits possible. The area was surrounded by water, and the principal northern political boundary was marked dramatically and unmistakably by a river. The thirty-eighth parallel seems to have been a powerful focus for a stalemate; and the main alternative, the "waist," was a strong candidate not just because it provided a shorter defense line but because it would have been clear to both sides that an advance to the waist did not necessarily signal a determination to advance farther and that a retreat to the waist did not telegraph any intention to retreat farther.

The Formosan Straits made it possible to stabilize a line between the Communist and National government forces of China, not solely because water favored the defender and inhibited attack, but because an island is an integral unit and water is a conspicuous boundary. The sacrifice of any part of the island would have made the resulting line unstable; the retention of any part of the mainland would have been similarly unstable. Except at the water's edge, all movement is a matter of degree; an attack across water is a declaration that the "agreement" has been terminated.

In Korea, weapons were limited by the qualitative distinction between atomic and all other; it would surely have been much more difficult to stabilize a tacit acceptance of any limit on size of atomic weapons or selection of targets.[9] No definition of size or target is so obvious and natural that it goes without saying, except for "no size, on any target." American assistance to the French forces in Indochina was persuasively limited to material, not people; and it was appreciated that an enlargement to include, say, air participation could be recognized as limited to air, while it would not be possible to establish a limited *amount* of air or ground participation. One's intentions to abstain from ground intervention can be conveyed by the complete withholding of ground forces; one cannot nearly so easily commit *some* forces and communicate a persuasive limit to the *amount* that one intends to commit.

The strategy of retaliation is affected by the need to communicate or coordinate on limits. Local aggression defines a place;

[9] This point is developed at length in Appendix A.

with luck and natural boundaries, there may be tacit acceptance of geographical limits or limits on types of targets. One side or both may be willing to accept limited defeat rather than take the initiative in breaching the rules, and to act in a manner that reassures the other of such willingness. The "rules" may be respected because, if they are once broken, there is no assurance that any new ones can be found and jointly recognized in time to check the widening of the conflict. But if retaliation is left to the method and place of the retaliator's own choosing, it may be much more difficult to convey to the victim what the proposed limits are, so that he has a chance to accept them in his counter-retaliation. In fact, the initial departure of retaliation from the locality that provokes it may be a kind of declaration of independence that is not conducive to the creation of stable mutual expectations. Thus the problem of finding mutually recognized limits on war is doubly difficult if the definition implicit in the aggressor's own act is not tolerable.

In sum, the problem of limiting warfare involves not a continuous range of possibilities from most favorable to least favorable for either side; it is a lumpy, discrete world that is better able to recognize qualitative than quantitative differences, that is embarrassed by the multiplicity of choices, and that forces both sides to accept some dictation from the elements themselves. The writer suggests that the same is true of restrained competition in every field in which it occurs.

PRIOR ARRANGEMENTS

While the main burden of this paper has been that tacit bargaining is possible and is susceptible of systematic analysis, there is no assurance that it will succeed in any particular case or that, when it succeeds, it will yield to either party a particularly favorable outcome compared with alternatives that might have been available if full communication had been allowed. There is no assurance that the next war, if it comes, will find mutually observed limits in time and of a sort to afford protection, unless explicit negotiation can take place. There is reason, therefore, to consider what steps can be taken before the time for tacit bargaining occurs, to enhance the likelihood of a successful outcome.

Keeping communication channels open seems to be one obvious point. (At a minimum, this might mean assuring that a surrender offer could be heard and responded to by either side.) The technical side of this principle would be identification of who would send and receive messages, upon what authority, over what facilities, using what intermediaries if intermediaries were used, and who stood in line to do the job in what fashion if the indicated parties and facilities were destroyed. In the event of an effort to fight a restrained nuclear war, there may be only a brief and busy instant in which each side must decide whether limited war is in full swing or full war has just begun; and twelve hours' confusion over how to make contact might spoil some of the chances for stabilizing the action within limits.

Thought should be given to the possible usefulness of mediators or referees. To settle on influential mediators usually requires some prior understanding, or at least a precedent or a tradition or a sign of welcome. Even if we rule out overt arrangements for the contingency, evidences by each side of an appreciation of the role of referees and mediators, even a little practice in their use, might help to prepare an instrument of the most extreme value in an awful contingency.

But all such efforts may suffer from the unwillingness of an adversary to engage in any preparatory steps. Not only may an adversary balk at giving signs of eagerness to come to agreement; it is even possible that one side in a potential war may have a tactical interest in keeping that war unrestrained and aggravating the likelihood of mutual destruction in case it comes. Why? Because of the strategy of threats, bluffs, and deterrents. The willingness to start a war or take steps that may lead to war, whether aggression or retaliation to aggression, may depend on the confidence with which a nation's leaders think a war could be kept within limits. To be specific, the willingness of America to retaliate against local aggression with atomic attack depends — and the Russians know that it depends — on how likely we consider it that such retaliation could itself remain limited. That is, it depends on how likely it is in our judgment that we and the Russians, when we both desperately need to recognize limits within which either of us is willing to lose the war without enlarging those limits, will find such limits and come to mutually recognized acquiescence in them. If, then, Russian refusal to engage in any activity that might lead to the possibility of limited war deters our own resolution to act, they might risk forgoing such limits for the sake of reducing the threat of American action. One parachutist in our example may know that the other will be careless with the plane if he is sure they can meet and save themselves; so if the first abstains from discussing the contingency, the other will have to ride quietly for fear of precipitating a fatal separation in the terrain below.

Whether this consideration or just the usual inhibitions on serious negotiation make prior discussion impossible, there is still a useful idea that emerges from one of our earlier games. It is that negotiation or communication for the purpose of coordinating expectations need not be reciprocal: unilateral negotiation may provide the coordination that will save both parties. Furthermore, even an unwilling member cannot necessarily make himself unavailable for the receipt of messages. Recall the man who proposed the letter R in one of the bargaining games: as long as the partner heard — and it is obvious that he heard — the letter R is the only extant proposal, and, being unchallenged, it may coordinate in default of any counterproposal nearly as well as if it had been explicitly accepted. (Even *denial* of it by the other party might not manage to dislodge its claim to prominence but rather simply prove his awareness of it, as long as no rival claim was made that created ambiguousness.) If one of our parachutists, just before the plane failed and while neither of them dreamed of having to jump, idly said, "If I ever had to meet somebody down there, I'd just head for the highest hill in sight," the other would probably recall and know that the first would be sure he recalled and would go there, even though it had been on the tip of his tongue to say, "How stupid," or "Not me, climbing hurts my legs," when the plane failed. When some signal is desperately needed by *both* parties and both parties know it, even a poor signal and a discriminatory one may command recognition, in default of any other. Once the contingency is upon them, their interests, which originally diverged in the play of threats and deterrents, substantially coincide in the desperate need for a focus of agreement.

PART II

A REORIENTATION OF
GAME THEORY

4

TOWARD A THEORY OF
INTERDEPENDENT DECISION

On the strategy of pure conflict — the zero-sum games — *game theory* has yielded important insight and advice. But on the strategy of action where conflict is mixed with mutual dependence — the nonzero-sum games involved in wars and threats of war, strikes, negotiations, criminal deterrence, class war, race war, price war, and blackmail; maneuvering in a bureaucracy or in a traffic jam; and the coercion of one's own children — traditional game theory has not yielded comparable insight or advice. These are the "games" in which, though the element of conflict provides the dramatic interest, mutual dependence is part of the logical structure and demands some kind of collaboration or mutual accommodation — tacit, if not explicit — even if only in the avoidance of mutual disaster. These are also games in which, though secrecy may play a strategic role, there is some essential need for the signaling of intentions and the meeting of minds. Finally, they are games in which what one player *can* do to avert mutual damage affects what another player *will* do to avert it, so that it is not always an advantage to possess initiative, knowledge, or freedom of choice.

Traditional game theory has, for the most part, applied to these mutual-dependence games (nonzero-sum games) the methods and concepts that proved successful in studying the strategy of pure conflict. The present chapter and the one to follow attempt to enlarge the scope of game theory, taking the zero-sum game to be a limiting case rather than a point of departure. The proposed extension of the theory will be mainly along two lines. One is to identify the perceptual and suggestive element in the formation of mutually consistent expectations. The other (in the following chapter) is to identify some of the basic "moves" that may occur in actual games of strategy, and the structural elements that the moves depend on; it involves such concepts as "threat," "enforcement," and the capacity to communicate or to destroy communication.

That game theory is underdeveloped along these two lines may reflect its preoccupation with the zero-sum game. Suggestions and inferences, threats and promises, are of no consequence in the accepted theory of zero-sum games. They are of no consequence because they imply a relation between the two players that, unless perfectly innocuous, must be to the disadvantage of one player; and he can destroy it by adopting a minimax strategy, based, if necessary, on a randomizing mechanism. So the "rational strategies" pursued by two players in a situation of pure conflict — as typified by pursuit and evasion — should not be expected to reveal what kind of behavior is conducive to mutual accommodation, or how mutual dependence can be exploited for unilateral gain.

If the zero-sum game is the limiting case of pure conflict, what is the other extreme? It must be the "pure-collaboration" game in which the players win or lose together, having identical preferences regarding the outcome. Whether they win fixed shares of the total or shares that vary with the joint total, they must rank all possible outcomes identically, in their separate preference scales. (And, to avoid any initial conflict, it has to be evident to the players that the preferences are identical, so that there is no conflict of interest in the information or misinformation that they try to convey to each other.)

What is there about pure collaboration that relates it to game theory or to bargaining? A partial answer, just to establish that this game is not trivial, is that it may contain problems of perception and communication of a kind that quite generally occur in nonzero-sum games. Whenever the communication structure does not permit players to divide the task ahead of time according to an explicit plan, it may not be easy to coordinate behavior in the course of the game. Players have to understand each other, to discover patterns of individual behavior that make each player's actions predictable to the other; they have to test each other for a shared sense of pattern or regularity and to exploit clichés, conventions, and impromptu codes for signaling their intentions and responding to each other's signals. They must communicate by hint and by suggestive behavior. Two vehicles trying to avoid collision, two people dancing together to unfamiliar music, or members of a guerrilla force that become separated in combat have to concert their intentions in this fashion, as do the applauding members of a concert audience, who must at some point "agree" on whether to press for an encore or taper off together.

If *chess* is the standard example of a zero-sum game, *charades* may typify the game of pure coordination; if *pursuit* epitomizes the zero-sum game, *rendezvous* may do the same for the coordination game.

An experiment of O. K. Moore and M. I. Berkowitz provides a nice mixture in which the two limiting cases are both visible.[1] It involves a zero-sum game between two teams, each team consisting of three people. The three members of the team have identical interests but, because of a special feature of the game, cannot behave as a single entity. The special feature is that the three members of each team are separated and can communicate only by telephone and that all six telephones are connected on the same line so that everyone can hear both the other team and his own teammates. No prearrangement of codes is permitted. Between teams we have here a pure-conflict game; among the members of the team we have a pure-coordination game.

If in this game we suppress the "other team" and if the three players simply try to coordinate a winning strategy in a game of skill or chance in the face of communication difficulty, we have a three-person pure-coordination game. Several "games" of this sort have been studied, both experimentally and formally; in fact, there is substantial overlap at this point between the nonzero-sum game and organization or communication theory.[2]

[1] O. K. Moore and M. I. Berkowitz, *Game Theory and Social Interaction*, Office of Naval Research, Technical Report, Contract No. SAR/NONR–609 (16) (New Haven, November, 1956).

[2] An extensive formal analysis of the coordination problem is developed by Jacob Marschak, "Elements for a Theory of Teams," and, "Toward an Eco-

The experiments reported in Chapter 3 showed that coordinated choice is possible even in the complete absence of communication. Further, they showed that there are tacit bargaining situations in which the *conflict* of interest in the choice of action may be overwhelmed by the sheer need for concerting on *some* action; in those situations, the limiting case of pure coordination isolates the essential feature of the corresponding nonzero-sum game.

So we do have, in this *coordinated problem-solving*, with its dependence on the conveyance and perception of intentions or plans, a phenomenon that brings out an essential aspect of the nonzero-sum game; and it stands in much the same relation to it as the zero-sum game, namely, that of "limiting case." One is the mixed conflict-cooperation game with all scope for cooperation eliminated; the other is the mixed conflict-cooperation game with the conflict eliminated. In one the premium is on secrecy, in the other on revelation.

It is to be stressed that the pure-coordination game is a *game of strategy* in the strict technical sense. It is a behavior situation in which each player's best choice of action depends on the action he expects the other to take, which he knows depends, in turn, on the other's expectations of his own. This interdependence of expectations is precisely what distinguishes a game of strategy from a game of chance or a game of skill. In the pure-coordination game the interests are convergent; in the pure-conflict game the interests are divergent; but in neither case can a choice of action be made wisely without regard to the dependence of the outcome on the mutual expectations of the players.[3]

---

nomic Theory of Organization and Information," *Cowles Foundation Discussion Papers*, Nos. 94 and 95 (New Series), and, with Roy Radner, "Structural and Operational Communication Problems in Teams," *Cowles Foundation Discussion Papers, Economics*, No. 2076. Examples of relevant empirical work can be found in Alex Bavelas, "Communication Patterns in Task-oriented Groups," in D. Cartwright and A. F. Zander, *Group Dynamics* (Evanston, 1953), G. A. Heise and G. A. Miller, "Problem Solving by Small Groups Using Various Communication Nets," in P. A. Hare, E. F. Borgatta, and R. F. Bales, *Small Groups* (New York, 1955), H. J. Leavitt and R. A. H. Mueller, "Some Effects of Feedback on Communication," in *Small Groups*, and L. Carmichael, H. P. Hogan, and A. A. Walter, "An Experimental Study of the Effects of Language on the Reproduction of Visually Perceived Form," *Journal of Experimental Psychology*, 15:73–86 (February, 1932).

[3] Concerning this point, Carl Kaysen in his review of Von Neumann and

Recall the famous case of Holmes and Moriarty on separate trains, neither directly in touch with the other, each having to choose whether to get off at the next station. We can consider three kinds of payoff. In one, Holmes wins a prize if they get off at different stations, Moriarty wins it if they get off at the same station; this is the zero-sum game, in which the preferences of the two players are perfectly correlated inversely. In the second case, Holmes and Moriarty will both be rewarded if they succeed in getting off at the same station, whatever station that may be; this is the pure-coordination game, in which the preferences of the players are perfectly correlated positively. The third payoff would show Holmes and Moriarty both being rewarded if they succeed in getting off at the same station, but Holmes gaining more if both he and Moriarty get off at one particular station, Moriarty gaining more if both get off at some other particular station, both losing unless they get off at the same station. This is the usual nonzero-sum game, or "imperfect-correlation-of-preferences" game. This is the mixture of conflict and mutual dependence that epitomizes bargaining situations. By specifying particular communication and intelligence systems for the players, we can enrich the game or make it trivial or provide an advantage to one of the two players in the first and third variants.

The essential game-of-strategy element is present in all three cases: the best choice for either depends on what he expects the other to do, knowing that the other is similarly guided, so that each is aware that each must try to guess what the second guesses the first will guess the second to guess and so on, in the familiar spiral of reciprocal expectations.

Morgenstern's *Theory of Games and Economic Behavior* says: "The theory of such games of strategy deals precisely with the actions of several agents, in a situation in which all actions are interdependent, and where, in general, there is no possibility of what we called parametrization that would enable each agent (player) to behave as if the actions of the others were given. In fact, it is this very lack of parametrization which is the essence of a game." Similar language is used by R. Duncan Luce and Howard Raiffa in *Games and Decisions* (New York, 1957): "Intuitively, the problem of conflict of interest is, for each participant, a problem of individual decision making under a mixture of risk and uncertainty, the uncertainty arising from his ignorance as to what the others will do" (p. 14). Their preoccupation is with the conflict, however; the case of coincident preferences they dispose of as trivial (pp. 59, 88), and they deal with such players as a single individual (p. 13).

A RECLASSIFICATION OF GAMES

Before going further, we can usefully reclassify game situations. The twofold division into zero-sum and nonzero-sum lacks the symmetry that we need and fails to identify the limiting case that stands opposite to the zero-sum game. The essentials of a classification scheme for a two-person game could be represented on a two-dimensional diagram. The values of any particular outcome of the game, for the two players, would be represented by the two coordinates of a point. All possible outcomes of a pure-conflict game would be represented by some or all of the points on a negatively inclined line, those of a pure common-interest game by some or all of the points on a positively inclined line. In the mixed game, or bargaining situation, at least one pair of points would denote a negative slope and at least one pair a positive slope.[4]

[4] If the nature of the game makes it desirable for a player to use a random device in the choice of his strategy, or feasible for the players to negotiate an enforcible agreement that, like a drawing of lots, depends on a chance mechanism, there may be room for cooperation in the choice of *strategies* even when there is perfect disagreement over the ranking of *outcomes*. In that case the points representing the pure-conflict game must meet the tighter restriction of lying on a straight line, with the two axes measuring the players' "utilities" in the sense now familiar in game theory. This restriction also applies to the pure common-interest game, since players who agree perfectly on the ranking of *outcomes* may not agree on the desirability of, say, one particular point over a fifty-fifty chance between the two points immediately above and below it. Thus "strictly pure" conflict and common-interest games, providing no scope for collaboration in the one case and no scope for disagreement in the other, would have to show the *expected values* of all pertinent mixed (random) strategies lying along the downward-sloping and upward-sloping lines, respectively, with axes measured in "utility units" of the kind mentioned; this in turn means that the points denoting *outcomes* must lie on a *straight* line.

Also, the pure games cannot admit "side payments." If one of the partners in a pure common-interest game threatens to sabotage the effect unless he is paid — assuming that the communication and enforcement structure of the game makes this possible — a conflict of interest is introduced; in effect, the point denoting the payment of a bribe would appear to the upper left or lower right of another point or points on the upward-sloping line, producing the configuration of a mixed game. And if one of the players in a pure-conflict game can threaten damage or offer compensation to induce his opponent to yield in this game, there is scope for bargaining; there is no longer a relation of pure conflict, and the points denoting the threatened damage or promised compensation would lie off the downward-sloping line. In other words, *all* pertinent potential outcomes must be allowed for. (Two simultaneous pure-conflict games, even if

We could stay close to traditional terminology, with respect to the strictly pure games, by calling them *fixed-sum* and *fixed-proportions* games, getting the unwieldy *variable-sum–variable-proportions* as the name for all games except the limiting cases. We could also call them perfect-negative-correlation games and perfect-positive-correlation games, referring to the correlation of their preferences with respect to outcomes, leaving for the richer mixed game the rather dull title of "imperfect-correlation game."

The difficulty is in finding a sufficiently rich name for the mixed game in which there is both conflict and mutual dependence. It is interesting that we have no very good word for the *relation* between the players: in the common-interest game we can refer to them as "partners" and in the pure-conflict game as "opponents" or "adversaries"; but the mixed relation that is involved in wars, strikes, negotiations, and so forth, requires a more ambivalent term.[5] In the rest of this paper I shall refer to the mixed game as a *bargaining game* or *mixed-motive game*, since these terms seem to catch the spirit. "Mixed-motive" refers not, of course, to an individual's lack of clarity about his own preferences but rather to the ambivalence of his relation to the other player — the mixture of mutual dependence and conflict, of partnership and competition. "Nonzero-sum" refers to the mixed game together with the pure common-interest game. And, because it characterizes the problem and the activity involved, *coordination game* seems a good name for the perfect sharing of interests.

GAMES OF COORDINATION

While most of this book will be about the mixed game, a brief discussion of the pure coordination game, beyond that of Chapter

---

they meet the restriction of straight lines, provide room for negotiation unless the slopes of the two lines happen to be identical.)

[5] It deserves to be emphasized that nonzero-sum games can as properly be classed under theory of partnership as under theory of conflict; and for providing insight into problems like that of limiting war, there is merit in using words that bring out the common interest of the adversaries and the "bargaining process" involved in the military maneuvers themselves. As will be seen in Chapter 9, even the problem of surprise attack is logically equivalent to a problem in partnership discipline. If *theory of games* has become endowed with a too conflict-oriented connotation, perhaps something like *theory of interdependent decision* would be a neutral term that equally covers the two limiting cases as well as the mixed case.

3, will help to show that this is an important game in its own right and will identify certain qualities of the mixed game that appear most clearly in the limiting case of pure coordination. Recall the various pure coordination problems of Chapter 3. Each of them evidently provided some focal point for a concerted choice, some clue to coordination, some rationale for the convergence of the participants' mutual expectations. It was argued there that the same kind of coordinating clue might be a potent force not only in pure coordination but in the mixed situation that includes conflict; and, in fact, the experiments demonstrated that, in the complete absence of communication, this is certainly true. But there are a number of instances in which pure coordination itself — the *tacit* procedure of identifying partners and concerting plans with them — is a significant phenomenon. A good example is the formation of riotous mobs.

It is usually the essence of mob formation that the potential members have to know not only where and when to meet but just when to act so that they act in concert. Overt leadership solves the problem; but leadership can often be identified and eliminated by the authority trying to prevent mob action. In this case the mob's problem is to act in unison without overt leadership, to find some common signal that makes everyone confident that, if he acts on it, he will not be acting alone. The role of "incidents" can thus be seen as a coordinating role; it is a substitute for overt leadership and communication. Without something like an incident, it may be difficult to get action at all, since immunity requires that all know when to act together. Similarly, the city that provides no "obvious" central point or dramatic site may be one in which mobs find it difficult to congregate spontaneously; there is no place so "obvious" that it is evident to everyone that it is obvious to everyone else. Bandwagon behavior, in the selection of leadership or in voting behavior, may also depend on "mutually perceived" signals, when a part of each person's preference is a desire to be in a majority or, at least, to see some majority coalesce.[6]

Excessively polarized behavior may be the unhappy result of

[6] A closely related phenomenon is appreciated by the person who tries to blend into the crowd to avoid being called on to recite, picked on by a bully, or singled out for "election" to some post that everybody wants to escape.

dependence on tacit coordination and maneuver. When whites and Negroes see that an area will "inevitably" become occupied exclusively by Negroes, the "inevitability" is a feature of convergent expectation.[7] What is most directly perceived as inevitable is not the final result but the *expectation* of it, which, in turn, makes the result inevitable. Everyone expects everyone else to expect everyone else to expect the result; and everyone is powerless to deny it. There is no stable focal point except at the extremes. Nobody can expect the tacit process to stop at 10, 30, or 60 per cent; no *particular* percentage commands agreement or provides a rallying point. If tradition suggests 100 per cent, tradition could be contradicted only by explicit agreement; if coordination has to be tacit, compromise may be impossible. People are at the mercy of a faulty communication system that makes it easy to "agree" (tacitly) to move but impossible to agree to stay. Quota systems in housing developments, schools, and so forth, can be viewed as efforts to substitute an explicit game with communication and enforcement for a tacit game that has an undesirably extreme "solution."

The coordination game probably lies behind the stability of institutions and traditions and perhaps the phenomenon of leadership itself. Among the possible sets of rules that might govern a conflict, tradition points to the particular set that everyone can expect everyone else to be conscious of as a conspicuous candidate for adoption; it wins by default over those that cannot readily be identified by tacit consent. The force of many rules of etiquette and social restraint, including some (like the rule against ending a sentence with a preposition) that have been divested of their relevance or authority, seems to depend on their having become "solutions" to a coordination game: everyone expects everyone to expect everyone to expect observance, so that nonobservance carries the pain of conspicuousness. Clothing styles and motorcar fads may also reflect a game in which people do not wish to be left out of any majority that forms and are not

[7] The phenomenon, called "tipping," is analyzed by M. Grodzins, "Metropolitan Segregation," *Scientific American*, 197:33–41 (October, 1957). A more innocuous example of explosively convergent expectations, based on tacit communication that has an almost electric quality, is the snicker that ignites an outburst of uncontrollable laughter in a nervous crowd. An important example was the collapse of the Batista regime, or of the Fourth Republic.

organized to keep majorities from forming. The concept of *role* in sociology, which explicitly involves the expectations that others have about one's behavior, as well as one's expectations about how others will behave toward him, can in part be interpreted in terms of the stability of "convergent expectations," of the same type that are involved in the coordination game. One is trapped in a particular role, or by another's role, because it is the only role that in the circumstances can be identified by a process of tacit consent.

A good example might be the *esprit de corps* (or lack of it) of an army unit or naval vessel or the value system of a particular college or fraternity. These are social organisms that are subject to a substantial rate of replacement but that maintain their own peculiar identities to an extent that does not seem to be accounted for by selective or biased recruitment. The individual character of one of these units seems to be largely a matter of convergent expectations — everyone's expectation of what everyone expects of everyone — with the new arrivals' expectations being molded in time to help mold the expectations of subsequent arrivals. There is a sense of "social contract," the particular terms of which are sensed and accepted by each incoming generation. I am told that this persistence of a tradition in a social entity is one of the reasons why the legal identity of an army division or regiment — its name and number and history — is often deliberately preserved when its strength has fallen to where abolition might seem indicated: the tradition that goes with the legal identity of the group is an asset worth preserving for a future buildup. It may be the same phenomenon that makes it possible to collect income tax in some countries and not in others: if appropriate mutual expectations exist, people will expect evasion to be on a scale small enough not to overwhelm the authorities and may consequently pay up either out of a sense of reciprocated honesty or out of fear of apprehension, thus together justifying their own expectations.

Nature of the intellectual process in coordination. It should be emphasized that coordination is not a matter of guessing what the "average man" will do. One is not, in tacit coordination, trying to guess what another will do in an objective situation; one is trying to guess what the other will guess one's self to guess the other to guess, and so on ad infinitum. ("Meeting" someone in the personal column of a newspaper is a good example.[8]) The reasoning becomes disconnected from the objective situation, except insofar as the objective situation may provide some clue for a concerted choice. The analogy is not just trying to vote with the majority but trying to vote with a majority when everyone wants to be in a

[8] So is meeting on the same radio frequency with whoever may be signaling to us from outer space. "At what frequency shall we look? A long spectrum search for a weak signal of unknown frequency is difficult. But, just in the most favored radio region there lies a unique, objective standard of frequency, which must be known to every observer in the universe: the outstanding radio emission line at 1420 megacycles of neutral hydrogen" (Giuseppe Cocconi and Philip Morrison, *Nature*, Sept. 19, 1959, pp. 844–846). The reasoning is amplified by John Lear: "Any astronomer on earth would say, 'Why, 1420 megacycles of course! That's the characteristic radio emission line of neutral hydrogen. Hydrogen being the most plentiful element beyond the earth, our neighbors would expect it to be looked for even by tyros in astronomy' " ("The Search for Intelligent Life on Other Planets," *Saturday Review*, Jan. 2, 1960, pp. 39–43). What signal to look for? Cocconi and Morrison suggest a sequence of small prime numbers of pulses, or simple arithmetic sums.

And this suggests an alternative orientation of those experiments in which subjects are instructed to make guesses, throughout a long random sequence of red or green lights, whether red or green will come up next. Subjects apparently persist in guessing on the basis of some pattern they think they perceive, an "irrational" mode of behavior given their knowledge that the sequence is generated by a random device. But, as Herbert Simon points out, "Man is not only a learning animal; he is a pattern-finding and concept-forming animal" ("Theories of Decision-Making in Economics and Behavioral Science," *American Economic Review*, 44:272). Why not, then, add to the experiment a cooperating pattern-maker, who generates the signals subject to various constraints and random interferences, and let the persistent pattern-seeking subject use his skill in finding the pattern planted by a cooperative partner rather than spend it futilely on random series? If, to make it tax the communicators' ingenuity, we add a third party whose reward is inversely related to that of the cooperating partners, who is allowed to intercept the message and within limits to alter it, we have something akin to the game of Moore and Berkowitz described earlier. Enriching the materials available beyond the binary choice of red and green might provide scope for genuinely creative pattern forming, of the kind that is interesting for Gestalt psychology, esthetics, and even higher-order problem solving. Simon notes in the same article (p. 426) that even a computer can be programed "to use something akin to imagery or metaphor in planning its proofs" of geometrical theorems. This is pattern seeking of real interest. (It reminds us that the assumption of "malevolent nature" by the zero-sum game theorist is not applicable to, say, mathematical invention. Nature gives hints; she presents her secrets in patterns that make them infinitely easier to guess than if an exhaustive scanning were required to find them.)

majority and everyone knows it — not to predict Miss Rheingold of 1960 but to buy the stock or real estate that everyone expects everyone to expect everyone to buy. Investment in diamonds may be a perfect example; the greatest of all may be the monetary role of gold, which can perhaps be explained only as the "solution" of a coordination game. (A common household version of the coordination game occurs when two people are cut off in a telephone conversation; if they both call back, they only get busy signals.)

Consider the game of "name a positive number." Experiments like those of Chapter 3 demonstrate that most people, asked just to pick a number, will pick numbers like 3, 7, 13, 100, and 1. But when asked to pick the same number the others will pick when the others are equally interested in picking the same number, and everyone knows that everyone else is trying, the motivation is different. The preponderant choice is the number 1. And there seems to be good logic in this: there is no unique "favored number"; the variety of candidates like 3, 7, and so forth, is embarrassingly large, and there is no good way of picking the "most favorite" or most conspicuous. If one then asks what number, among all positive numbers, is most clearly unique, or *what rule of selection would lead to unambiguous results,* one may be struck with the fact that the universe of all positive numbers has a "first" or "smallest" number.[9]

---

[9] There is a widely quoted passage in Keynes (p. 156) that may be worth repeating in order to point out that, while it deals with exactly the problem dealt with here, its conception of the "solution" is *not* at all the same: "Professional investment may be likened to those newspaper competitions in which the competitors have to pick out the six prettiest faces from a hundred photographs, the prize being awarded to the competitor whose choice most nearly corresponds to the average preference of the competitors as a whole; so that each competitor has to pick, not those faces which he himself finds prettiest, but those which he thinks likeliest to catch the fancy of the other competitors, all of whom are looking at the problem from the same point of view. It is not a case of choosing those which, to the best of one's judgment, are really the prettiest, nor even those which average opinion genuinely thinks prettiest. We have reached the third degree where we devote our intelligence to anticipating what average opinion expects the average opinion to be. And there are some, I believe, who practice the fourth, fifth, and higher degrees" (J. M. Keynes, *The General Theory of Employment, Interest and Money* [New York, 1936], p. 156). This class of games demonstrates, incidentally, that the usual correlation between parametric behavior and large numbers does not hold for tacit

Game-theory formulation of the coordination problem. The pay-off matrix for a pure coordination problem would look something like that in Fig. 8. One player chooses a row, the other a column;

1	0	0	0	0
0	1	0	0	0
0	0	1	0	0
0	0	0	1	0
0	0	0	0	1

FIG. 8

and they receive the rewards denoted by the numbers contained in the cell where their choices intersect. If to each choice of one player there corresponds a single choice for the other that "wins" for both of them, we can arrange columns so that all the winning cells lie along the diagonal. In those cells there are positive pay-offs to both players, in the rest we can put zeros. (For our present purpose there is nothing lost by letting a single number stand in each cell for the payoff to both players.)

But we must rule out a possible axiom that might seem to be suggested by analogy with other game theories, namely, that (to use the term of Luce and Raiffa) the "labeling" of rows, columns, and players should make no difference to the outcome.[10] It is pre-

---

play with multiple equilibria. To adapt "parametrically" to the behavior of others requires in this case that their behavior be observable, not conjectural; the nonparametric character of tacit coordination remains, no matter how large the number of players.

[10] Labeling of the *players* is explicitly ruled out by Luce and Raiffa (pp. 123–127) in discussing cooperative games and in effect is ruled out by Nash in his symmetry assumption (J. F. Nash, "The Bargaining Problem," *Econometrica*, 18:155–162 [1950], and "Two Person Cooperative Games," *Econometrica*, 21:128–140 [1953]). Labeling of *strategies* for tacit or explicit nonzero-sum games is implicitly precluded by dealing only with games in normal form, that is, the abstract version of them as represented by a payoff matrix (which is itself an *analytical* device, not part of the game, and hence provides no left-right, upper-lower, or numerical ordering of the actual strategies). A good example in which the labeling of *players* is the controlling factor is the interrupted

cisely because strategies are "labeled" in some sense — that is, have symbolic or connotative characteristics that transcend the mathematical structure of the game — that players can rise above sheer chance and "win" these games; and it is for that same reason that these games are interesting and important.

Even the game portrayed in Fig. 8 which might seem to have a minimum of symbolic significance attached to rows and columns, is not a hard one to "win," that is, for players to do substantially better on than chance would suggest, if it is portrayed in a matrix as shown. (If we give that same game an infinite series of rows and columns, it seems to become easier rather than harder. In that case it is formally identical with the game mentioned earlier, "Pick a positive number," but, because the "labeling" is different, there is less tendency for minorities to congregate at 3, 7, 13, and so forth.) Just forming the matrix prejudices the choice, since it focuses attention on "first," "middle," "last," and so forth.[11] If strategies are not given sequential labels, that is, labels that can be ordered like numbers and alphabets, but are given individual names, and these are not presented in any particular order, it is the names that must coordinate choice.

And here it becomes emphatically clear that the intellectual processes of choosing a strategy in pure conflict and choosing a strategy of coordination are of wholly different sorts. At least this is so if one admits the "minimax" solution, randomized if necessary, in the zero-sum game. In the pure-coordination game, the player's objective is to make contact with the other player through some imaginative process of introspection, of searching for shared clues; in the minimax strategy of a zero-sum game — most strikingly so with randomized choice — one's whole objective is to avoid any meeting of minds, even an inadvertent one.[12]

---

telephone call mentioned earlier, with the problem of who should call back and who should wait for the call.

[11] This point is typical of a number of demonstrations in the author's experiments reported earlier, to the effect that the postulate regarding the "independence of irrelevant alternatives" cannot be credited in the tacit game and, for analogous reasons, should not be expected to hold in the explicit bargaining game. Potential outcomes can be relevant to the coordination of choice, though not themselves near to being chosen. For a statement and discussion of this postulate see Luce and Raiffa, p. 127.

[12] Randomized strategies may nevertheless be useful to achieve a coordinated

To illustrate, suppose that I am to name one card in an ordinary deck of fifty-two and you are to guess which one I name. Traditional game theory gives guidance on how to make my choice on the assumption that I do not want you to outguess me; I can select at random and defy you to have a better than random chance of guessing what I name. But if the game is that I *do* want you to guess correctly and you know that I will try to pick one that facilitates your guess, the random device can only guarantee to make tacit cooperation impossible. Holmes can *destroy* the labeling of the stations by flipping a coin to decide where to get off the train; and Moriarty has only a fifty-fifty chance of guessing a coin. But in the common-interest version they must somehow *use* the labeling of the stations in order to do better than pure chance; and how to use it may depend more on imagination than on logic, more on poetry or humor than on mathematics. It is noteworthy that traditional game theory does not assign a "value" to this game: how well people can concert in this fashion is something that, though hopefully amenable to systematic analysis,

---

distribution of votes, say, among a panel of candidates. If a 55 per cent majority exists and knows that it does, among a hundred voters; if two out of six candidates are congenial to it; and if the three candidates polling the largest numbers of votes become the board of directors, there is danger that uncoordinated polling may concentrate too many votes on the first (or second) majority choice, leaving the minority two winning candidates with 22 votes apiece. But if each member of the majority flips a coin to cast his vote for one of his party's men, the likelihood of one's getting as few as 22 votes is only one chance in six. If the minority, too, lacks an overt means of collaborating and relies on a chance device, the majority's chances are excellent.

A partial randomized strategy may also be used to reduce an area of conflict. Suppose two people, seated at North and East sides of a card table, are to move to another card table adjacent that is identically oriented, must choose without communication what seats they will take at the other table, and will win prizes of $1 apiece if they pick adjacent seats. This is an easy coordination problem; but let us subvert the incentives, by giving an additional $2 premium to the player who is on the other's right in the event they succeed in sitting next to each other. This game has no equilibrium point; interests do not converge; there is no seating arrangement that would not give one an incentive to move. (Each may wish that he could promise to sit on the other's left, but cannot.) A random strategy yields each player a minimax value of $1. But, if each decides where he would sit in the pure common-interest game, then flips a coin to see whether he does sit there or sits opposite, the players guarantee that they neither choose the same seat nor sit opposite each other and share equal chances of winning the premium. This is an equilibrium pair of (mixed) strategies, worth an expected value of $2 apiece.

cannot be discovered by reasoning a priori. This corner of game theory is *inherently* dependent on empirical evidence.[13]

It should particularly be noted that to assert the influence of "labels" (that is, of the symbolic and connotative details of the game) and the dependence of the theory on empirical evidence does not involve the question of whether game theory is predictive or normative — concerned with generalizations about actual choice or the strategy of correct choice. The assertion here is *not* that people simply *are* affected by symbolic details but that they *should* be for the purpose of correct play. A normative theory must produce strategies that are at least as good as what people can do without them. More, it must not deny or expunge details of the game that can demonstrably benefit two or more players and that the players, consequently, should not expunge or ignore in their mutual interest. Two couples jockeying for space on a dance floor or two armies jockeying for a truce line may jointly suffer

[13] In cases like this we need only to consider the question of what *price* players would pay for a bit of coordinating information, and what different information patterns yield what chances of coordinating, to find ourselves in the middle of Marschak's *theory of teams*.

There is, incidentally, a version of "prisoners' dilemma" for this game: two accomplices, apprehended before their alibi is prepared and interrogated separately, must concert the alibi they invent or be revealed in their guilt. A tantalizing variant can be built by supposing that confession carries a lighter sentence than unconfessed guilt; each player has a "minimax" strategy of confession and must not only consider which particular alibi constitutes the *best* alibi strategy but *how good it is* (in terms of likely coincidence with his partner's) and whether they share the decision to try it. The matrix might be:

.5	.5	o	.5	o	.5	o
o	.5	1		o		o
o	.5		o	1		o
o	.5		o		o	1

(Lower left entry in each cell is payoff to player choosing row, upper right to player choosing column.)

from decision processes that are limited to the abstract properties of the situation.

A particular implication of this general point is that the game in "normal" (mathematically abstract) form is not logically equivalent to the game in "extensive" (particular) form, once we admit the logic by which rational players concert their expectations of each other. As pointed out in Chapter 3, these same considerations seem to be powerfully present in explicit bargaining as well. A terminological implication of these considerations is that "noncooperative" is a poor name for the game of tacit coordination; it is desperately cooperative in its own peculiar way and is still so when we add conflict and form the tacit mixed-motive game (In Appendix C it is argued that certain solution concepts familiar in game theory can be given an interpretation in terms of the coordination concept.)

SUGGESTION AND MUTUAL PERCEPTION IN THE MIXED-MOTIVE GAME

Coordination-game theory, while interesting in its own right, is interesting mainly for the light that it sheds on the nature of the mixed-motive game. The coordination element shows up most strikingly in a purely tacit game, in which there is neither communication nor any sequence of moves by which the two players accommodate themselves to each other. An example, similar to problem 6 on page 62, would be the following.

One player is "located" in Cincinnati, the other in San Francisco; they have identical maps of the United States and are to divide the country between them. Each is to draw a line dividing the United States into two parts; the line may be straight or curved, related or unrelated to physical or political landmarks. If the two of them divide the map differently, neither gets anything; but if they draw identical division lines on their maps, they are both rewarded. The reward for each player depends on what is contained in his piece after the division, that is, the piece that contains the city in which he is located. Let us leave these rewards vague; they may depend partly on area, partly on population, partly on industrial wealth and agricultural resources, and so forth, and may differ somewhat for the two players. In other words, while all terrain is valuable, not all parts of the country are equally valuable, and there is no clear specification of the valuation formula. (There is consequently no means of selecting a perfectly symmetrical division of values between the two players.)

In this game there is a compelling problem of coordination; each player can win only if he does exactly what the other expects him to, knowing that the other is similarly trying to do exactly what is expected of him. They must jointly find a line that in some fashion suggests itself to both of them or appeals to both of them. Neither can "outsmart" the other without outsmarting himself.

The experiments of Chapter 3 suggest that players are by no means helpless when faced with this kind of game. The game is nowhere near so "infinitely" difficult as the infinity of possible division lines might suggest; some variants of the game are not difficult at all. But a successful outcome does depend on the kinds of factors that are controlling in the pure-coordination game; in fact, some games of this sort are "won" by the two players' choosing exactly the same outcome as they would have chosen if the reward system gave them identical, instead of conflicting, interests. The problem is to find some signal or clue or rationalization that both can perceive as the "right" one, with each party prepared to be disciplined by that signal or clue in the event that it appears to discriminate against him. They must find their clues where they can. (If the map they are using happens, for example, to contain an embarrassing richness of clues, making it difficult to single out any particular one, a fairly arbitrary line drawn as a suggestion by the referee, identical on both maps, might have to be accepted as a "mediator," even if it is substantially biased toward one of the players.)

But this coordination element, especially in the case without conflict, appears to be essentially related to a *communication problem*. The pure-coordination game not only ceases to be interesting but virtually ceases to be a "game" if the players can concert with certainty, without difficulty, and without cost. The question arises, then, how important the coordination element can be in mixed-motive games generally, since many of these take the form of overt bargaining with uninhibited speech.

The pervasiveness of the coordination principle arises from two separate considerations. One, which was discussed in Chapter 3 is that tacit bargaining provides an analytical model — perhaps only an analogy but perhaps an identification of the actual psychic and intellectual phenomenon — of the "rational" process of finding agreement in pure bargaining situations, those in which both parties recognize that there is a wide range of outcomes preferable to both of them over no agreement at all. The psychic phenomenon of "mutual perception" that can be verified as real and important in the tacit case has a role to play in the analysis of explicit bargaining. *Coordination of expectations* is the role.

Second, many of the bargaining processes or game situations that we want to analyze are at least partly tacit. In some cases, like maneuvering a car in a traffic jam, speech is physically precluded; in others, like developing a modus vivendi with a neighbor, speech is inhibited in the interests of privacy. Illicit bargaining, or diplomatic bargaining that would be embarrassing to both sides if overheard by other countries, may be less than fully articulate. If the number of players in a game is large, as it is in the bargaining process that determines the racial border lines between residential areas and professions, there may be no institutional provision for explicit negotiation. In these cases, while speech may be part of the bargaining process, actions are also part of it, and the game is one of "maneuver" rather than just talk.

Furthermore, if there are *moves* available to the players, so that it is an advantage to get on with the maneuver even while negotiating, and particularly if some maneuvers become visible to the other player only after a time lag, there is no reason to suppose that an instantaneous moratorium on maneuver will reign from the outset; in that case, the game progresses while the talk is going on. If the moves had only symbolic significance, we could include them in the communication process along with speech; but, typically, moves have a tactical significance, leaving the game irreversibly different from what it was before, and typically also their tactical significance raises them above the level of pure speech even in their communication content. One may say and say that a gun is loaded without being able to prove it until he actually shoots; one may say and say that he considers an area strategically important and not be believed until he incurs expense or risk in its protection. Thus moves can reveal information about a player's value system or about the choices of action available to him; moves can commit him to certain actions when speech often cannot; and moves can often progress at a speed that is determined unilaterally, not dependent on formalities of agreement at a conference.

In other words, bargaining games quite typically involve a dynamic process of mutual accommodation rather than pure communication culminating in a crystallized agreement. The jockeying for limits in limited war is a perfect example, and we might illustrate it by modifying the parlor game described above.

An illustrative tacit game. Suppose our two players with their maps of the United States before them are each given 100 chips and told to play a game as follows.[14] At each "move," each player will distribute five chips among states on his map. The moves are compared, and if the two players have put a chip apiece in the same state, those two chips are removed; if one player has put a chip and the other player three chips in the same state, a chip apiece is removed leaving only two chips representing the one player; and so forth. They do the same at the next move, again with five chips; this time they have the option of placing their chips on states that are yet uncovered or of placing them on states where there are already chips. If A puts two chips on a state in

[14] Since it will be proposed in Chapter 6 that such games have, in fact, a research value, as well as an illustrative value, it should be observed at the outset that there is a special problem of motivating the players in an experimental nonzero-sum game. In a zero-sum game, winning is measured relative to one's immediate adversary, and the intellectual challenge and bilateral competition motivate the player toward the correct (and only) type of winning. But for a mixed-motive game, "winning" must be made to involve one's absolute score, not his score relative to that of the person he plays with; the incentives are distorted if the play is dominated by strictly bilateral competition. So, unless real rewards are given, the game has to be organized as a round robin or some such schedule that involves more than two players in a series of two-person plays, with the final outcome decided by the relative position of one's absolute score. (This is why there are no two-person nonzero-sum parlor games.)

which B previously put a chip, B's is removed along with one of A's, leaving one of A's chips present to "claim" the state. And so the game goes until the players have used up all their chips; it then continues, and at each move a player may transfer up to five chips from the states in which they are to other states, again with equal numbers of chips being removed from a state in which both players have placed chips. This process goes on until both players have notified the referee that they are willing to terminate the game.

Prizes are now distributed. Each player receives a dollar for every one of his chips still on the board, that is, for those that were not removed when he "took" a state or "lost" it to the other player. He also gets money for the states that he "possesses," these being the states that he has chips on plus those without chips that are in the area containing his home base that is completely inclosed by states that he does have chips on.

These "rewards" for states possessed are specific dollar values attached to each of the 48 states; they vaguely follow a pattern suggestive of, say, "economic worth" or something of the sort. There is no presumption that the values are the same, or even very closely correlated, for the two players; population may be an important element in the "values" of the states for one of the players and a comparatively unimportant element in the "values" for the other player. Neither player knows the other player's value system — or perhaps knows just a little about it, such as what elements matter but not how much they matter. Each must learn what he can about the other's value system by observing the other player's moves.

Here we have a mixed-motive game, which progresses by a process of mutual accommodation — a series of moves in the course of which the players suffer damage jointly if their accommodation is poor. They may lose dollars by failing to predict where each other will place his chips during the current move, in those cases where they prefer not to lose dollars fighting over a state. Each loses at least a dollar when one takes a state from the other; and they may lose more than a dollar apiece if the one who loses a state attempts to recapture it by putting more chips on it. And not only do they lose a dollar with each dollar forfeited, but each player has fewer "chips" left from the point of view of claiming states; and they may have to leave some states completely unclaimed between them if they have not enough chips left on the board when the game ends.

Now how do the players "bargain" in this game? One way or another, they do in fact make proposals and counterproposals; they accept, reject, retaliate, and even discover ways of conveying threats and promises.[15] But if we deny them any form of speech, they must convey their intentions and their proposals by their patterns of behavior. Each must be alert to what the other is expressing in his maneuvers, and each must be inventive enough to convey his intentions when he wants them conveyed. If one player badly wants a particular state, because it has especially high value for him, so that he is willing to stick around and fight it out a long time, losing several dollars to the kitty before the other player gives up, it is better for both players that they realize ahead of time which one wants it most badly. And if a player is really prepared to concede a large portion of the country as a "trade" for some other portion that he badly wants, he must not only make it conspicuously available to the other side but must somehow demarcate its limits by his own pattern of play.

But where do the patterns come from? They are not very richly provided by the mathematical structure of the game, particularly since we have purposely made each player's value system too uncertain to the other to make considerations of symmetry, equality, and so forth, of any great help. Presumably, they find their patterns in such things as natural boundaries, familiar political groupings, the economic characteristics of states that might enter their value systems, Gestalt psychology, and any clichés or traditions that they can work out for themselves in the process of play.[16]

[15] This has been evident in preliminary experiments with such a game.

[16] If my neighbor's fruit tree overhangs my yard and I pick exactly all the fruit on my side of the line, my neighbor can probably discern what my "proposal" is, and has a good idea of what he has acquiesced in for the future if he does not retaliate. But if, instead, I pick that same amount of fruit from both sides of the line haphazardly or pick some amount that is related, say, to the size of my family, he is less likely to perceive just what I have in mind. (He may also be more obliged to resist or retaliate if I pick only *part* of the fruit on my side of the line than if I pick it all, since I have failed to demarcate the limit of my intentions.)

Explicit communication. Now let us change the rules so that the players may talk as much as they please. How different would this make the game? In some respects, it should increase the efficiency of the players; particular trades can be identified now that were too complex to make proposals about under the more clumsy system. Perhaps, too, the players can avoid some of the inadvertent clashes of chips on the same state, which cost them dollars. We cannot be sure that they will avoid mutually costly competitive bidding for states, since the advantage of being first on a state is great enough to motivate players to keep playing even while they talk. And they have no way to persuade each other that they mean what they say except by showing it in the way they play. (We let them tell each other how they value the states; but we explicitly make fibs unpunishable, and we provide the players no written evidence of their value systems that they could show each other.)

So the introduction of uninhibited speech may not greatly alter the character of the game, even though the particular outcome is different. The dependence of the two players on conveying their intentions to each other and perceiving the intentions of each other, of behaving in predictable patterns and acquiescing in rules or limits, is much the same as before.

The contrast with a zero-sum game and the peculiarly self-effacing quality of a minimax solution is striking here. With a minimax solution, a zero-sum game is reduced to a completely unilateral affair. One not only does not need to communicate with his opponent, he does not even need to know who the opponent is or whether there is one. A randomized strategy is dramatically anticommunicative; it is a deliberate means of destroying any possibility of communication, especially communication of intentions, inadvertent or otherwise. It is a means of expunging from the game all details except the mathematical structure of the payoff, and from the players all communicative relations.

In chess it does not matter whether the pieces look like horses, ecclesiastics, elephants, castles, or hamburger buns; whether the game is called "chess," "civil war," or "real estate"; or whether the squares are distorted to look like political or geographical subdivisions. It does not matter what the players know about each other or whether they speak the same language and have a common culture; nor does it matter who played the game previously and how it came out. (If it did matter, one of the players would be motivated to destroy the influence of these details; and a minimax strategy, randomized if necessary, would destroy it.)

But change the payoff matrix in a chess game, making it a non-zero-sum game that rewards the players not only for the pieces they capture but for the pieces they have left over at the end, as well as the squares they occupy, in such fashion that both players have some interest in minimizing the "gross" capture of pieces with its mutual destruction of value. Make each player uncertain about just what squares and what particular pieces the other player values most. And have moves by the clock, so that neither player can hold up the other player's moves for the sake of talking to him.

Now it may make a difference to the players whether we call the game "war" or "gold rush"; whether the pieces look like horses, soldiers, explorers, or children on an Easter egg hunt; what map or picture is superimposed on the playing board and how the squares are distorted into different shapes; or what background story the players are told before they begin.

We have now rigged the game so that the players must *bargain* their way to an outcome, either vocally or by the successive moves that they make, or both. They must find ways of regulating their behavior, communicating their intentions, letting themselves be led to some meeting of minds, tacit or explicit, to avoid mutual destruction of potential gains. The "incidental details" may facilitate the players' discovery of expressive behavior patterns; and the extent to which the *symbolic* contents of the game — the suggestions and connotations — suggest compromises, limits, and regulations should be expected to make a difference. It should, because it can be a help to both players not to limit themselves to the abstract structure of the game in their search for stable, mutually nondestructive, recognizable patterns of movement. The fundamental psychic and intellectual process is that of participating in the creation of *traditions;* and the ingredients out of which traditions can be created, or the materials in which potential traditions can be perceived and jointly recognized, are not at all coincident with the mathematical contents of the game.[17]

The outcome is determined by the expectations that each player forms of how the other will play, where each of them knows that their expectations are substantially reciprocal. The players must jointly discover and mutually acquiesce in an outcome or in a mode of play that makes the outcome determinate. They must together find "rules of the game" or together suffer the consequences.

A good example of this problem of communicating intentions is that of getting across, persuasively, an intended pattern of retaliation for particular acts that one proposes to consider "out of bounds." Without full communication, one's ability to convey such a pattern of intentions is dependent not only on the contextual materials available for the formation of bounds and limits but on the capacity of the other player to recognize the formula (Gestalt) of retaliation when he sees a sample of it. Historical and literary precedent, legal and moral casuistry, mathematics and aesthetics, as well as familiar analogues from other walks of life, may constitute the menu from which one has to choose his recognizable pattern of retaliation as well as his interpretation of the other's intended pattern. Even with full verbal communication, the situation may not be greatly different; patterns of action may speak louder than words.

Thus the influence that the suggestive details of a game may have on its outcome and the dependence of the players on what

[17] A good example is the question whether a clear line can be drawn between atomic and other weapons, the answer to which is reported now to be negative if explosive power is the criterion, the explosive ranges having overlapped. But there is nevertheless a difference if enough people think so, and they undoubtedly do. It is a difference constructed of the pure fabric of expectations: it is a ten years' *tradition* that atomic weapons *are* different; people believe so and believe others to believe so, and even those who deny the difference will undoubtedly catch their breath, whenever the next one goes off in a war, in a manner they cannot explain by reference to the force of the explosion. It is a purely conventional difference, like the one that makes imprisonment not a "cruel and unusual" punishment or that makes, say, university representation in Parliament perfectly compatible with English democracy if it has always existed but not if it has to be reinstated after a ten years' lapse. The atomic-weapons difference is also one that, probably, can be deliberately reinforced or deliberately blurred over time, as most traditions can. (This point is developed at length in Appendix A.)

clues and signals the game provides are relevant not merely to the study of how players actually do behave in a nonzero-sum game. It is not being argued that players just *do* respond to the non-mathematical properties of the game but that they *ought* to take them into account, hence that even a normative theory — a theory of the *strategy* of games — must recognize that rational players may jointly take advantage of them. And even when one rational player realizes that the configuration of these details discriminates against him, he may also rationally recognize that he has no recourse — that the other player will rationally expect him to submit to the discipline of the suggestions that emanate from the game's concrete details and will take actions that, on pain of mutual damage, assume he will co-operate.[18]

[18] It should be added that the concept of the intrinsic magnetism or focusing quality of particular outcomes in a bargaining situation or in a pure coordination problem gets some support and clarification from the very substantial body of experimental evidence provided by the Gestalt psychologists. Their work on the perception of physical forms is pertinent. For example, incomplete shapes were shown to people whose vision was damaged in part of the eye, and they often saw the shapes as complete rather than as partial. But the particular shapes that they "completed" for themselves followed certain principles of simplicity; and unfamiliar "simple" figures were completed where very familiar, but less simple, figures were not. Koffka refers to "spontaneous organization in simple shapes." We are surrounded by skewed rectangles; but what we "see" about us is rectangles, not departures from perfect rectangles, because "the true rectangle is a better organized figure than the slightly inaccurate one would be." Adverting to the minimum-maximum properties of stationary processes, Koffka suggests that psychological processes will have these properties: "For we can at least select psychological organizations which occur under simple conditions and can then predict that they must possess regularity, symmetry, simplicity. This conclusion is based on the principle of isomorphism, according to which characteristics of the physiological processes are also characteristic aspects of the corresponding conscious processes." And, "Thus we have gained a general, though admittedly somewhat vague, principle to guide us in our investigation of psychophysical organization. . . . The principle . . . can briefly be formulated like this: psychological organization will always be as 'good' as the prevailing conditions allow. In this definition the term 'good' is undefined. It embraces such properties as regularity, symmetry, simplicity and others which we shall meet in the course of our discussion" (K. Koffka, *Principles of Gestalt Psychology* [London, 1955]).

If individual perception and "organization" of forms follow these constraints, the process of "mutual perception" and "mutual organization of forms" involved in the convergence of expectations must depend on similar restraints at least as rigorous. And, since the nonzero-sum game requires some ultimate joint "organization of form," so to speak, a normative theory of strategy (not just a descriptive psychology) must take these restraints into account.

A hypothetical experiment. As an illustration of what the author has in mind, the following hypothetical experiment can be considered. (Hopefully, some such experiment could be carried out.) It is offered here as a conceptual analogue or, conceivably, an empirical test of the psychic phenomenon involved in bargaining.

The first stage in the experiment is to invent a machine, perhaps on the principle of the lie detector, that will record or measure a person's "recognition" or the focus of his attention or his alertness or his excitement. What we want is a machine that measures, as the player scans an array of possible outcomes in some orderly fashion, the extent to which particular outcomes catch his attention or generate excitement in the course of actual bargaining.

Given the machine, set up a bargaining game. For simplicity, make it one in which there are certain gains to be shared when agreement is reached on the shares. Give the game enough "topical content" to provide some room for argument, casuistry, alternative rationales, and so forth; that is, provide more than a bare mathematical range with a conspicuous mid-point.

Now have the two players connected to their machines in such a way that each can see the meter on his own machine, each can see the meter on the other's machine, and each is aware that both are aware that both can see both meters. In other words, they mutually perceive that they both can see each other's reactions to particular outcomes as they come within view of the scanning device. We employ a mechanical scanning device, which moves about in the range of possible outcomes, pointing to, lighting up, or focusing on one possible outcome after another. It follows perhaps some regular course, perhaps a random course. Let this machine scan; let the players watch it scan, watch their own and each other's meters, and watch each other's faces if they wish to.

Finally, we go through with the game; and there may be several variants. An interesting possibility would be to exclude explicit bargaining and simply let the scanning proceed, back and forth or round and round among the array of alternative outcomes. We watch to see whether the recorded reactions of the two players tend eventually to converge on a single outcome, in the sense that their involuntary, physically identifiable reactions are at some kind of maximum for the same particular outcome among all those to which the scanning device elicits their reactions. (For control purposes, we might once have subjected each player to a scanning session in which the other player was absent, to get some notion of each player's reactions independently of any interaction between the players.) If convergence does occur, we have certainly identified a significant phenomenon, whether or not we can allege that this is *the* psychic bargaining process. We shall have demonstrated (*a*) that players do react to the content of the bargaining situation and (*b*) that their reactions are subject to a mutual interaction that results from the fact that each can see the other's reaction and each knows that his own visible reaction is yielding information about his own expectations. (The writer conjectures that, like Lot's wife, players will often be unable to keep their attention from being drawn to particular outcomes, even unfavorable outcomes, and that a conscious effort to ignore a "focal point" may often enhance the focal power.) [19]

Another variant would be to let the players bargain explicitly during the scanning and metering, with the scanning device inexorably eliciting their physical reactions in the course of the discussion in a manner visible to both of them. (We could even, in this latter case, let a player adduce the evidence of the visible reaction meters if he wished to as a bargaining tactic, pointing out to his partner, for example, that the latter "obviously" cannot expect to hold out for, say, the $60 he is verbally demanding when it is clear from his blood pressure that his mind is settled on $40.)

This experiment would rest on three hypotheses. First, that an individual player would have physically identifiable "reactions" upon contemplating different alternatives among the range of

[19] The following observation, quoted by Koffka, may be hard to believe but is certainly to the point: "When an expert . . . follows a football game attentively he will also notice that the goalkeeper, standing before the comparatively large goal, is more often hit than can be accounted for by the mere adventitious kicking of the contestants, even when one takes account of the fact that the goalkeeper whenever he can will try to intercept the ball. The goalkeeper furnishes a prominent point in space which attracts the eyes of the opposing kickers. If the motor activity takes place while the kicker's eye is fixed on the goalkeeper, then the ball will generally land near him. But when the kicker learns to reconstruct his field, to change the phenomenal 'centre of gravity' from the goalkeeper to another point in space, the new centre of gravity will have the same attraction as the goalkeeper had before."

possible game outcomes and that these reactions would be conspicuously different among the different alternatives. Second, that these reactions, when the player knows that they are naked to his partner's eye, would behave in a manner suggestive of bargaining; that is, that the reactions of the two players, when visible to both of them, would interact in a kind of "bargaining process." Third, that this measured phenomenon, which we liken to a bargaining process, is part of, or is involved in, or is related to, *the* bargaining process as defined in the ordinary way. (An experiment of the sort described might prove especially interesting for the case of more than two persons.)

The experiment has not been carried out and is not adduced as evidence. It has been described here in order to give an operational representation of the theoretical system that the author has in mind in referring to the "convergence" of expectations and to suggest that the convergence that ultimately occurs in a bargaining process may depend on the dynamics of the process itself and not solely on the a priori data of the game.

Some dynamic characteristics of focal-point solutions. The dependence of a "focal-point" solution on some characteristic that distinguishes it qualitatively from the surrounding alternatives has important dynamic considerations. For example, it often makes small concessions less likely than large ones; it often means that the focal point is more persuasive as an *exact* expected outcome than as an approximation. If a bargainer has persistently been unsuccessfully demanding 50 per cent, compromise at 47 per cent is unlikely; the small concession may be a sign of collapse. Qualitative principles are hard to compromise, and focal points generally depend on qualitative principles. One cannot expect to satisfy an aggressor by letting him have a few square miles on this side of a boundary; he knows that we both know that we both expect our side to retreat until we find some persuasive new boundary that can be rationalized.

In fact, a focal point for agreement often owes its focal character to the fact that small concessions would be impossible, that small encroachments would lead to more and larger ones. One draws a line at some conspicuous boundary or rests his case on some conspicuous principle that is supported mainly by the rhetorical question, "If not here, where?" The more it is clear that concession is collapse, the more convincing the focal point is. The same point is illustrated in the game that we play against ourselves when we try to give up cigarettes or liquor. "Just one little drink," is a notoriously unstable compromise offer; and more people give up cigarettes altogether than manage to reach a stable compromise at a small daily quota. Once the virgin principle is gone, there is no confidence in any resting point, and expectations converge on complete collapse. The very recognition of this keeps attention focused on the point of complete abstinence.

Sometimes the focal point itself is inherently unstable. In that case it serves not as an outcome but as a sign of where to look for the outcome. This is often true of a "test vote" in a legislative body or a "test issue" that arises in the relations between the players in some continuing game. Often it is a challenge or a dare or an act of defiance that, by its nature, must either elicit a submissive response from the other party or be submissively withdrawn. It is a small piece of the game that comes to symbolize the game itself, setting a pattern of expectations that extends beyond the substance of the point involved. Sometimes it is so intended and constitutes a deliberate tactic; in other cases the act or the issue develops an unintended symbolic significance, making compromise impossible.

Diplomatic recognition of the Communist regime in China, loyalty oaths at universities, a strike settlement in a key industry, surrender of the floor to an interrupter at a cocktail party, or the vote on some particular motion at a political convention may all have this kind of significance. Sometimes, it is true, the outcome on this particular issue simply yields evidence of how other issues would be decided, as when a test vote indicates exactly how large the opposition to a measure is; but often the particular issue is not representative of the rest of the game, it just acquires tacit recognition as a clue to all that will follow, so that each side is the prisoner or beneficiary of the mutual expectations that are created.

Often this phenomenon can be identified as an actual signal in a coordination game. The members of an unorganized coalition can often recognize the potentialities of concerted action without being sure that "agreement" exists to act in concert. One wants to know how everyone else is going to act and whether everyone else will do what he knows he ought to. A test vote in a legislature or some particular simultaneous action among the group, like a mass protest, is often a means of "ratifying" the existence of the coalition and of demonstrating that everybody expects everybody else to act in concert. But even in a two-person game, as typified by the dare, the phenomenon of psychological dominance or submissiveness may prove to be psychologically identical with the resolution of a bargaining game.

This process, by which particular moves in a game or offers and concessions achieve symbolic importance as indicators of where expectations should converge in the rest of the game, seems to be an area in which experimental psychology can contribute to game theory.

The Empirical Relevance of Mathematical Foci. We must avoid assuming that everything the analyst can perceive is perceived by the participants in a game, or that whatever exerts power of suggestion on the analyst does so on the participant in a game. In particular, game characteristics that are relevant to sophisticated mathematical solutions (except when the same solution can also be reached by an alternative, less sophisticated route) might not have this power of focusing expectations and influencing the outcome. They might have it only if the players perceived each other to be mathematicians. This may be the empirical interpretation of such "solutions" as those of Braithwaite, Nash, Harsanyi, and others. It is that the mathematical properties of a game, like the aesthetic properties, the historical properties, the legal and moral properties, the cultural properties, and all the other suggestive and connotative details, can serve to focus the expectations of certain participants on certain solutions. If two players are themselves mathematical game theorists, they may mutually perceive and be powerfully affected by potential solutions that have compelling mathematical properties. Each may transcend, and know that the other will transcend, various adventitious details that, to nonmathematician game players, might be more relevant to the focusing of expectations than some of the quantitative properties of the game.

(In many cases these mathematical properties would be a uniqueness or symmetry that would have nonmathematical definitions and nonmathematical appeal, too, or would happen to coincide with qualitatively distinguishable points that could be rationalized in an equally compelling nonmathematical way.)

Thus mathematical solutions are one species of a genus of influences that have the power to focus expectations; but they work through the same psychic mechanism — this power of suggestion that is able to bring expectations into convergence — as the other species. When husband and wife, separated in a department store, gaily traipse off to the Lost and Found by a tacit and jocular mutual appreciation that it is the "obvious" place to meet, two mathematicians in the same situation — each aware that both are aware that both are mathematicians — might look for a geometrically unique point rather than one that depended on a play on words.

The main point here is independent of whether, under the "rules" of game theory, a rational player must be presumed to know as much mathematics as he ever has need for. We are dealing here with the players' shared appreciations, preoccupations, obsessions, and sensitivities to suggestion, not with the resources that they can draw on when necessary. If the phenomenon of "rational agreement" is fundamentally psychic — convergence of expectations — there is no presumption that mathematical game theory is essential to the process of reaching agreement, hence no basis for presuming that mathematics is a main source of inspiration in the convergence process. (This topic is pursued further in Appendix B.)

One may or may not agree with any particular hypothesis how a bargainer's expectations are formed, either in the bargaining process or before it and either by the bargaining itself or by other forces. But it does seem clear that the outcome of a bargaining process is to be described most immediately, most straightforwardly, and most empirically in terms of some phenomenon of stabilized convergent expectations. Whether one agrees explicitly to a bargain or agrees tacitly or accepts it by default, he must, if he has his wits about him, expect that he could do no better and recognize that the other party must reciprocate the feeling. Thus the *fact* of an outcome, which is simply a coordinated choice, should be analytically characterized by the notion of converging expectations.

Communicating subjective information. The role of "expressive moves" in a mutual-accommodation game of this sort is enhanced by the consideration that in mixed-motive games, in contrast to zero-sum games that are known to the players to be zero-sum, there is likely to be uncertainty about each other's value system. Moves have an *information* content in the mixed-motive game.

Nor can we set up as a general case the bargaining game in which each side has foreknowledge of the other's preferences. To assume that either knows the "true" payoff matrix of the other is often to make an extraordinary assumption about the institutional arrangements of the game. The reason is that certain elements in a bargaining game are *inherently unknowable* for some of the participants, except when there are special conditions. How can we know how badly the Russians would dislike an all-out war in which both sides were annihilated? We cannot; and the reason we cannot is *not* solely that the Russians are necessarily unwilling that we should know. On the contrary, circumstances may arise in which they are desperate that we should know the truth. But how can they make us know it? How can they make us believe that what they tell us is true? How can the prisoner being tortured for secrets that he really does not know persuade his captors that he does not know them? How could the Chinese, if they were really determined to take Formosa at the cost of an all-out war, persuade us that they could not be deterred in any fashion and that any threat on our part would only commit us both to all-out war? [20]

[20] The lack of any means of testing the truth is the very basis of that tantalizing game in which each participant attaches positive value to the other's welfare, as when husband and wife discuss whether or not to go to a movie, each wanting to do whatever the other wants to do and wanting to seem to want it himself, knowing that the other is similarly expressing a preference

In special cases the information can be conveyed. In an artificial game, in which each player's "value system" is contained on cards or chips, he may simply turn them face up (if the rules permit or if he and his adversary can jointly cheat against the referee). In a society that believes absolutely in a superior power that will punish falsehood when asked to do so and that everybody knows everbody else believes in, "cross my heart and hope to die" is a sufficient formula for conveying truth voluntarily. But these are special cases. If we are to have a "general case" it must be one in which there is at least some ignorance of each other's value system, or each other's strategy options, if only because such facts are inherently unknowable or incommunicable.

Von Neumann and Morgenstern illustrated their *solution* concept for the nonzero-sum game with the example of a seller, A, prepared to sell his house for any price above 10, and two buyers, B and C, prepared to pay up to 15 and 25, respectively.[21] (My numbers.) The novel part of the solution was that C might pay B a share of his saving if, through B's staying out of the market, C got the house for less than 15. They proposed — and this limitation was inherent in their concept of *solution* — that the most B might receive from C was $15 - 10 = 5$. What is interesting about the information requirement of this solution is not that B's reservation price of 15 is something that he might try to misrepresent, but that in the ordinary world he could not convincingly communicate the truth if he wanted to. Not only does the "solution" concept — by its assumption of full information — rule out the intrusion of speculators (unless they genuinely want the house enough to give them a basis for sharing in the solution), but it assumes that C can discern, or B can reveal, a subjective truth,

---

that represents a guess at what one wants to do, etc. There is also an entire domain of game theory involving interpersonal relations in which the overt revelation or recognition of one's value system itself affects values; my awareness that my neighbor does not like me may cause me small discomfort, as does his awareness of my awareness, but if we are forced to accredit the fact overtly, the pain may be acute. "Social etiquette," remarks Erving Goffman, "warns men against asking for New Year's Eve dates too early in the season, lest the girl find it difficult to provide a gentle excuse for refusing." "On Face-Work," *Psychiatry: Journal for the Study of Interpersonal Processes*, 18:224 (1955).
    [21] J. Von Neumann and O. Morgenstern, *Theory of Games and Economic Behavior* (Princeton, 1953), pp. 564ff.

one that D and E (speculators who are attracted by the observation that B makes a pure bargaining profit in connection with an object that he never owns before or after) cannot counterfeit.

There are undoubtedly special cases in which one can suppose that the other player is like one's self in basic values and can consequently estimate the other's values by the simple application of symmetry. But in too many exciting cases one plays an opponent who is a wholly different kind of person. The father of a kidnapped boy will not be very successful in guessing what his own bottom price would be if he had been the kidnapper instead; it may not be easy for a British or French officer introspectively to guess how terrible a penalty would have to be to deter him if he were a Mau Mau or an Algerian terrorist. It is hard for a boy to guess how much he would like himself if he were the girl that he wants to date, or for the customer in the restaurant to know how much he would dislike a scene if he were the waiter instead.

This is one of the reasons why talk is not a substitute for moves. Moves can in some way alter the game, by incurring manifest costs, risks, or a reduced range of subsequent choice; they have an information content, or *evidence* content, of a different character from that of speech. Talk can be cheap when moves are not (except for the "talk" that takes the form of *enforcible* threats, promises, commitments, and so forth, and that is to be analyzed under the heading of *moves* rather than communication anyway). Mutual accommodation ultimately requires, if the outcome is to be efficient, that the division of gains be in accordance with "comparative advantage"; that is, the things a player concedes should be those that he wants less than the other player, relative to the things he trades for. Each needs, therefore, to communicate his value system with some truth, although each can also gain by deceiving. While one's maneuvers are not unambiguous in their revelation of one's value systems and may even be deliberately deceptive, they nevertheless have an evidential quality that mere speech has not.

The uncertainty that can usually be presumed to exist about each other's value systems also reduces the usefulness of the concept of mathematical *symmetry* as a normative or predictive principle. Mathematical symmetry cannot be perceived if one has access to only half the relevant magnitudes. To the extent that symmetry is helpful to the players in accommodating their movements to each other's, it would tend to be symmetry of a more qualitative sort, of the kind that depends on visible context rather than underlying values.

5

ENFORCEMENT, COMMUNICATION, AND STRATEGIC MOVES

Whenever we speak of deterrence, atomic blackmail, the balance of terror, or an open-skies arrangement to reduce the fear of surprise attack; when we characterize American troops in Europe as a trip wire or plate-glass window or propose that a threatened enemy be provided a face-saving exit; when we advert to the impotence of a threat that is so enormous that the threatener would obviously shrink from carrying it out or observe that taxi drivers are given a wide berth because they are known to be indifferent to dents and scratches, we are evidently deep in game theory. Yet formal game theory has contributed little to the clarification of these ideas. The author suggests that nonzero-sum game theory may have missed its most promising field by being pitched at too abstract a level of analysis. By abstracting from communication and enforcement systems and by treating perfect symmetry between players as the general case rather than a special one, game theory may have overshot the level at which the most fruitful work could be done and may have defined away some of the essential ingredients of typical nonzero-sum games. Preoccupied with the solution to *the* nonzero-sum game, game theory has not done justice to some typical game situations or game models and to the "moves" that are peculiar to nonzero-sum games of strategy.

What "model," for example, epitomizes the controversy over massive retaliation? What conditions are necessary for an efficacious threat? What in game theory corresponds to the proverbial situation "to have a bear by the tail"; how do we identify the payoff matrix, the communication system, and the enforcement system that it embodies? What are the tactics by which pedestrians intimidate automobile drivers, or small countries large ones; and how do we formulate them in game-theoretical terms? What is the information or communication structure, or the complex of incentives, that makes dogs, idiots, small children, fanatics, and martyrs immune to threats?

The precarious strategy of cold war and nuclear stalemate has often been expressed in game-type analogies: two enemies within reach of each other's poison arrows on opposite sides of a canyon, the poison so slow that either could shoot the other before he died;[1] a shepherd who has chased a wolf into a corner where it has no choice but to fight, the shepherd unwilling to turn his back on the beast; a pursuer armed only with a hand grenade who inadvertently gets too close to his victim and dares not use his weapon; two neighbors, each controlling dynamite in the other's basement, trying to find mutual security through some arrangement of electric switches and detonators.[2] If we can analyze the structures of these games and develop a working acquaintance with standard models, we may provide insight into real problems by the use of our theory.

To illustrate, an instructive model is that of twenty men held up for robbery or ransom by a single man who has a gun and six bullets. They can overwhelm him if they are willing to lose six of themselves, if they have a means of deciding which six to lose. They can defeat him without loss if they can visibly commit themselves to a threat to do so, if they can simultaneously commit themselves to a *promise* to abstain from capital punishment, once they have caught him. He can deter their threat if he can visibly commit himself to shoot in disregard of any subsequent threat they might make, or if he can show that he could not believe their promise. If they cannot deliver their threat — if, say, he understands only a foreign language — they cannot disarm him verbally. Nor can they make a threat unless they agree on it themselves; so if he can threaten to shoot any two who talk

---

[1] Compare C. W. Sherwin, "Securing Peace Through Military Technology," *Bulletin of the Atomic Scientists*, 12:159–164 (May 1956).

[2] Compare Herman Kahn and Erwin Mann, "Game Theory," The RAND Corporation, Paper P–1166 (Santa Monica, 1957), pp. 55ff. The authors work out a number of problems involving dynamite, detonators, and deterrence.

together, he can deter agreement. If the twenty cannot find a way to divide the risk, there may be no one to go first to carry out the threat, hence no way to make the threat persuasive; and if he can announce a formula for shooting, such as that those who move first get shot first, he can deter them unless they find a way to move together without a "first." If fourteen of the twenty can overpower the remaining six and force them to advance, they can demonstrate that they could overwhelm the man; if so, the threat succeeds and the gunman surrenders, and even the six "expendables" gain through their own inability to avoid jeopardy. If the twenty could overwhelm the man but have no way of letting him escape, a promise of immunity may be necessary; but if they cannot deny their capacity to identify him and testify against him later, it may be necessary to let him take a hostage. This, in turn, depends on the ability of nineteen to enforce their own agreement to protect, by silence, whoever is currently the hostage . . . and so on. When we have identified the critical ingredients in several games of this sort, we may be in a better position to understand the basis of power of an unpopular despot or of a well-organized dominant minority, or the conditions for successful mutiny.

This chapter is an attempt to suggest the kinds of typical moves and structural elements that deserve to be explored within the framework of game theory. They include such moves as "threat," "promise," "destruction of communication," "delegation of decision," and so forth, and such structural elements as the communication and enforcement provisions.

AN ILLUSTRATIVE MOVE

An example of a standard "move" is the commitment, analyzed at some length in Chapter 3. If the institutional environment makes it possible for a potential buyer to make a single "final" offer subject to extreme penalty in the event he should amend the offer — to commit himself — there remains but a single, well-determined decision for the seller: to sell at the price proposed or to forego the sale. The possibility of commitment converts an indeterminate bargaining situation into a two-move game; one player assumes a commitment, and the other makes a final decision. The game has become determinate.[3]

This particular move, analyzed at length in Chapter 3, is mentioned here only as a particularly simple illustration of a typical move. As noted in Chapter 3, the availability and the efficacy of this move depend on the communication structure of the game and the ability of the player to find a way to commit himself, to "enforce" the commitment against himself. Furthermore, we have allowed the move structure of the game to be asymmetrical; the "winner" is the one who can assume the commitment or, if both can, the one who can do it first. (We can consider the special case of a tie, but we have not, by an assumption of symmetry, made ties a foregone conclusion.)

But, although we have made the game "determinate" in the sense that we have no difficulty in identifying the "solution," once we have identified which of the two players can first commit himself, it remains a game of *strategy*. Though the winner is the one who achieves his commitment first, the game is not like a foot race that goes to the fastest. The difference is that the commitment does not automatically win under the rules of the game, either physically or legally. The outcome still depends on the second player, over whom the first player has no direct control. The commitment is a *strategic* move, a move that induces the other player to choose in one's favor. It constrains the other player's choice by affecting his expectations.

The power to commit one's self in this kind of game is equivalent to "first move." And if the institutional arrangements provide no means for incurring an irrevocable commitment

[3] In the real estate example of Von Neumann and Morgenstern referred to earlier (p. 116) buyer B (whose top price is 15) might raise the limit on what he can extract from buyer C (whose top price is 25) if he can find some means to bind himself to buy the house for 20 and keep or destroy it (that is, not be free to resell it to C for a loss) unless he gets a specified large fraction of, say, 20 — P, where P is the ultimate price paid by C. In effect, B changes his own "true" top price, thus raising the limit on what he may extract from C. Of course, D and E may try to do the same; and the first to get properly committed, or the one who can find a means if only one of them can, is the winner. If D, who attaches no personal value to the house, is committed to pay up to 22 for it, he is a bona fide member of the game with a true reservation price of 22; his *bona fides* is even greater than was B's originally, if the commitment is demonstrable while subjective valuations are not.

in a legal or contractual sense, one may accomplish the same thing by an irreversible maneuver that reduces his own freedom of choice. One escapes an undesired invitation by commitment when he arranges a "prior" engagement; failing that, he can deliberately catch cold. Luce and Raiffa have pointed out that the same tactic can be used by a person against himself when he wants, for example, to go on a diet but does not trust himself. "He announces his intention, or accepts a wager that he will not break his diet, so that later he will *not* be free to change his mind and to optimize his actions according to his tastes at *that* time." [4] The same thing is accomplished by maneuver rather than by commitment when one deliberately embarks on a vacation deep in the wilds without cigarettes.

THREATS

The distinctive character of a threat is that one asserts that he will do, in a contingency, what he would manifestly prefer not to do if the contingency occurred, the contingency being governed by the second party's behavior. Like the ordinary commitment, the threat is a surrender of choice, a renunciation of alternatives, that makes one worse off than he need be in the event the tactic fails; the threat and the commitment are both motivated by the possibility that a rational second player can be constrained by his knowledge that the first player has altered his own incentive structure. Like an ordinary commitment, a threat can constrain the other player only insofar as it carries to the other player at least some appearance of obligation; if I threaten to blow us both to bits unless you close the window, you know that I won't unless I have somehow managed to leave myself no choice in the matter.[5]

[4] *Games and Decisions*, p. 75.

[5] In ordinary language, "threat" is often used also for the case in which one merely points out to an adversary, or reminds him, that one would take action painful to the adversary if the latter fails to comply, it being clear that one would have incentive to do so. To "threaten" to call the police on a trespasser is of this sort, the threat to shoot him is not. But it seems better to use a different word for these cases — I suggest "warning" rather than "threat" — because the "threat" either is superfluous, and does not constitute a move, or it conveys true information and relates to situations with an information struc-

The threat differs from the ordinary commitment, however, in that it makes one's course of action *conditional* on what the other player does. While the commitment fixes one's course of action, the threat fixes a course of reaction, of response to the other player. The commitment is a means of gaining *first move* in a game in which first move carries an advantage; the threat is a commitment to a strategy for *second move*.

A threat can therefore be effective only if the game is one in which the first move is up to the other player or one can force the other player to move first. But if one must, in a mechanical sense, move first or simultaneously, he can still force the legal equivalent of "first move" on the other by attaching his threat to a demand that the other promise in advance how he will behave — if the game has communication and enforcement structures that make promises feasible and that the party to be threatened cannot destroy in advance. The holdup man whose rich victim happens to have no money on him at the time can make nothing of his opportunity unless he can extract a hostage while he awaits payment; and even that will not work unless he can himself find a way to assume a convincing commitment to return the hostage in a manner that does not subject himself to identification or capture.

The fact that *some* kind of commitment, or at least appearance of commitment, must lie behind the threat and be successfully communicated to the threatened party is in contradiction to another notion that often appears in game theory. This is the notion that a threat is desirable, or admissible, or plausible, only if the reaction threatened would cause worse damage to the threatened party than to the party making the threat. This is the view of

---

ture and communication structure worth keeping distinct. In this latter case it is a mutually beneficial move, precluding a jointly undesired outcome by improving the second party's understanding. The main point of analytical similarity, between this "warning" case and that of the "threat," is in the possible difficulty of conveying true information credibly, of conveying *evidence* for the assertion that one would have, ex post, incentive for doing as one warns he will. As a matter of fact, if a threat is of such nature (as it often is) that the act of commitment is not contained in the act of communicating it — if the commitment precedes the conveyance of the threat, with evidence for believing it, to the threatened party — the first act in the process of threatening changes the "true" incentive structure, and the second is, in effect, a "warning."

Luce and Raiffa, who characterize threats by the phrase, "This will hurt you more than it hurts me," explicitly making threats depend on interpersonal utility comparisons. In the event that both players attempt to make plausible threats, they say, the result becomes indeterminate, depending on the "bargaining personalities" of the players; "and to predict what will in fact happen without first having a complete psychological and economic analysis of the players seems foolish indeed." [6]

[6] Pp. 110–11, 119–20, 143–44. Morton A. Kaplan, in applying game theory to international relations, also takes the position that "any criterion giving weight to the threat positions of the players involves an interpersonal comparison of utilities." (See his *System and Process in International Politics* [New York, 1957].) Luce and Raiffa may partly be led to their view that only one of the players has a "plausible" threat to make, by confining their brief discussion to 2×2 matrices. It is impossible to show, with a 2×2 matrix, a game in which both players could be interested in making threats. A threat is essentially a credible declaration of a *conditional* choice for second move. It is profitable only if it yields a better payoff than either first move or second move alone and when one can make the other player move first either actually or by promise. (If second move alone is as good, the threat is unnecessary; and if first move were as good, one needs only an unconditional commitment to his strategy choice, not a commitment to a conditional choice.) But if this preference order holds for one player in a 2×2 matrix, it cannot hold for the other player. The actual matrices used by Luce and Raiffa in discussing the point show no "plausible" threat strategy for player No. 2, not because the absolute size of his gains or losses is greater than player 1's but for the much simpler reason that player 2 has no use for a threat. He wins if he moves first; he wins if he moves second; and he wins with simultaneous moves, in the games shown. His only interest in a threatlike declaration would be to forestall his partner's threat; and for that purpose he needs only an *unconditional* commitment to his preferred strategy — that is, the legal equivalent of "first move" in advance of his partner's threat. The "threat" tactic of J. F. Nash, which applies to bargaining games that have a continuous range of efficient outcomes — or that can be made to, by agreement on the odds in a drawing of lots — differs from the threat discussed here, in that the threatener does not demand, on pain of mutual damage, a *particular* outcome but only *some* outcome in the efficient range; that is, he shifts the zero point corresponding to "no agreement." The motive for that threat is the expectation of a particular mathematically determinate outcome whose locus is shifted by the shift in the payoffs corresponding to nonagreement. This is the kind of threat assumed by Luce and Raiffa (p. 139) in the "asymmetrical" game. The implicit legal structure of the game apparently honors no irrevocable commitments (otherwise, first commitment would easily win the game for either player). Each player is subject to the legal "disability" that he can always, by the overt act of explicit agreement with his partner on any outcome, evade his own commitment. This being so, the revocable commitments can only shift the zero point — the "status quo" that will rule unless explicit agreement on some outcome is reached. The "asymmetry" that is present in the particular

But the issue is both simpler and more precise than that. Consider the left-hand matrix in Fig. 9, where Column is assumed to have "first move." Without threats, Column has an easy "win." He chooses strategy I, forcing Row to choose between payoffs of 1 and 0; Row chooses strategy i, providing Column a payoff of 2. But if we allow Row to make a threat, he declares that he

	I	II
i	2 / 1	1 / 2
ii	0 / 0	0 / 0

	I	II
i	10 / 9	9 / 10
ii	8 / 0	0 / 0

FIG. 9

will choose strategy ii unless Column chooses II; that is, he gives Column a choice of ii,I or i,II by committing himself to that conditional choice. *If* Column went ahead and chose I, of course, Row would prefer to choose i; and they both know it. The tactic succeeds only if Column believes that Row *must* choose ii in the event of I.

Either he does believe this, or he does not. If he does not, the "threat" is nothing at all to him; he goes ahead and makes his "best" first move, choosing I. If he does believe that Row must follow a strategy of i,II or ii,I, Column prefers 1 to 0 and chooses II. But this is true of any numbers that we might put in the matrix that reflect the same order of preferences. It is true of the right-hand matrix as well. That one dramatizes the essential character of the threat more than the first one, since the penalty on Row of an irrational choice by Column is greater in this case; but for rational play and full information, Row need not worry. Column's preference is clear; and, once Row has given him the

game shown by Luce and Raiffa is thus a feature of the particular legal system that implicitly prevails. In practice it might correspond, say, to the deliberate incurring of social disapproval on failure to reach agreement, with such disapproval constituting cost or punishment (perhaps asymmetrical between participants) in addition to the cost of nonagreement but with the public not concerned with what the agreement provides as long as some agreement is reached.

pair to choose from — ii,I versus i,II — there is no doubt what Column will do. If I threaten to blow my brains all over your new suit unless you give me that last slice of toast, you'll give me the toast or not depending on whether you know that I've arranged to have to do so, exactly as if I'd only threatened to throw my scrambled eggs at you.[7]

The issue here is in whether or not we admit that the game has "moves," that is, that it is possible for one player or both players to take actions in the course of the game that irreversibly change the game itself — that in some fashion alter the payoff matrix, the order of choices, or the information structure of the game. If the game by its definition admits no moves of any sort, except mutual agreement and refusal to agree, then it may be true that the "personalities" of the players determine the outcome, in the sense that their expectations in a "moveless" game converge by a process that is wholly psychic. But, if a threat is anything more than an assertion that is intended to appeal to the other player by power of suggestion, we must ask what more it can be. And it must involve some notion of commitment — real or fake — if it is to be anything.

"Commitment" is to be interpreted broadly here. It includes maneuvers that leave one in such a position that the option of nonfulfilment no longer exists (as when one intimidates the other car by driving too fast to stop in time), maneuvers that shift the final decision beyond recall to another party whose incentive

[7] Edward Banfield showed me this irresistible quotation about the Bháts and Charáns of the west of India, revered as bards. "In Guzerát they carry large sums in bullion, through tracts where a strong escort would be insufficient to protect it. They are also guarantees of all agreements of chiefs among themselves, and even with the government.

"Their power is derived from the sanctity of their character and their desperate resolution. If a man carrying treasure is approached, he announces that he will commit trága, as it is called: or if an engagement is not complied with, he issues the same threat unless it is fulfilled. If he is not attended to, he proceeds to gash his limbs with a dagger, which, if all other means fail, he will plunge into his heart; or he will first strike off the head of his child; or different guarantees to the agreement will cast lots who is to be first beheaded by his companions. The disgrace of these proceedings, and the fear of having a bard's blood on their head, generally reduce the most obstinate to reason. Their fidelity is exemplary, and they never hesitate to sacrifice their lives to keep up an ascendency on which the importance of their cast depends" (The Hon. Mountstuart Elphinstone, *History of India* [ed. 7; London, 1889], p. 211).

structure would provide an ex post motive for fulfilment (as when the authority to punish is deliberately given to sadists, or when one shifts his claims and liabilities to an insurance company), and maneuvers that simply "worsen" one's own payoff in the contingency of nonfulfilment so that even the horror of a mutually damaging fulfilment becomes more attractive (as when one arranges for himself to appear a public coward if he fails to fulfil, or when he puts a plate-glass window in front of his wares or stations women and children on the particular bit of territory that he has threatened somewhat implausibly to defend at great cost). A nice everyday example is given by Erving Goffman, who reminds us that "salesmen, especially street 'stemmers,' know that if they take a line that will be discredited unless the reluctant customer buys, the customer may be trapped by considerateness and buy in order to save the face of the salesman and prevent what would ordinarily result in a scene." [8]

There are, however, some ways in which this notion of commitment to a threat can be usefully loosened. One is to recognize that "firm" commitment amounts to the invocation of some wholly potent penalty, such that one would in all circumstances prefer to carry out what he was committed to. It is a penalty of infinite (or at least of superfluous) size that one voluntarily, irreversibly, and visibly attaches to all patterns of action but the one that he is committed to do. This concept can be loosened by supposing that the penalty is of finite size and not necessarily so large as to be controlling in all cases. In Fig. 10 Column will

	I	II
i	2 / 4	1 / 1
ii	2 / 4	3 / 3

FIG. 10

---

[8] Goffman's paper is a brilliant study in the relation of game theory to gamesmanship and a pioneer illustration of the rich game-theoretic content of formalized behavior structures like etiquette, chivalry, diplomatic practice, and — by implication — the law.

win if he has first move, unless Row can commit himself to i. (Commitment obtains "first move" for Row.) But, if commitment means the attachment of a finite penalty to the choice of row ii and we show this in the matrix by subtracting from each of Row's payoffs in ii some finite amount representing the penalty, then the commitment will be effective only if the penalty is greater than 2. Otherwise it is clear to Column that Row's response to II will be ii, in spite of the commitment. In this case the commitment is simply a loss that Row would impose on himself, so he avoids it.

Similarly with a threat. In Fig. 11 without threats, the solution

	I	II	III
i	-5 $\quad -5$	-1 $\quad -2$	-1 $\quad -2$
ii	-3 -4	3 0	2 2
iii	-3 -4	1 1	0 3

FIG. 11

is at iii,II whether the rules call for Row to choose first, Column first, or both to choose simultaneously. Either player can win if he can move second and confront the other with a threat.[9] Column would threaten I against iii, Row would threaten i against

[9] If a player, Column, for example, cannot force first move on Row in a mechanical sense, he can do so in a "legal" sense by threatening to choose I unless Row *promises* to chose ii. Full analysis in this case requires attention to the penalties on promises as well as on threats. Since the physical and institutional arrangements for promises (that is, for commitments to the second party) are generally of a quite different nature from those for unilateral commitments (that is, commitments that the second player cannot himself dissolve), available penalties could differ drastically as between threats and promises — just as, in general, they would differ as between the first and second players. The particular payoffs shown in Fig. 4 would require penalties of at least 1 on a promise by Column or by Row. Note that in the case of a promise extracted by a threat, it is an advantage to the threatener to be able to invoke penalty and a disadvantage to the victim to be able to invoke penalty on his own breach of contract, that is, to be able to comply.

II. But if the threat is secured by a penalty, the lower limit to any persuasive penalty that Column could invoke would be 4; any smaller penalty leaves him preferring II to I when Row chooses iii. The lower limit to a persuasive penalty on Row's noncompliance would be 3. If, then, the situation is one in which penalties come in a single "size," a size less than 3 goes unused and the outcome is at iii,II; a size greater than 4 is adequate for either player, and the "winner" is the one who can avail himself of the threat first; a size between 3 and 4 is of use only to Row, who wins. In this latter case the player who would be hurt the more by his own unsuccessful threat is the one who cannot threaten — but only through the paradox that he is incapable of calling a sufficiently terrible penalty on his own head.

Note that the "hurt-more" comparison in this case refers not to whether Row or Column would be hurt more by what Row threatens but to whether Row would be hurt more by having to fulfil his own threat than Column would be hurt if, instead, Column had made *his* threat. Actually, in the particular payoff matrix shown, Row's *successful* threat is one that would hurt him *more* in the fulfilment than it would hurt Column, while Column's potential *unsuccessful* threat would hurt him *less* to fulfil than it would hurt Row.

Another loosening of the threat concept is to alter our assumption of rationality. Suppose there is some probability Pr for player R, and some probability Pc for player C, that he will make a mistake or an irrational move, or that he will act in an unanticipated way because the other player is mistaken about the first player's payoffs.[10] This yields us a game in which the possible gains and losses in committing one's self to a threat must take into account the possibility that a fully committed threat will not be heeded. If, then, the potential loss that will ensue from having to carry out the threat is greater for one player than for another, there could be symmetrical circumstances — the P's being equal and the threat penalties equal for the two players — in which one player may find it advantageous to make the threat and the other player not, considering the possibility of "error." (A somewhat similar calculation may be involved if both players

[10] Situations of this sort are explored in Chapters 7 and 9.

have opportunities for threats and there is danger of simultaneous commitment through the failure of one to observe the other's commitment and to stop in time to save both.)

This modification in the threat concept — in the rationality postulate that underlies it — goes somewhat in the direction of the "hurt-more" criterion. On the whole, though, game theory adds more insight into the strategy of bargaining by emphasizing the striking truth that the threat does *not* depend on the threatener's having less to suffer than the threatened party if the threat had to be carried out rather than by exaggerating the possible truth contained in the intuitive first impression. Threats of war, of price war, of damage suit; threats to make a "scene"; most of the threats of organized society to prosecute crimes and misdemeanors; and the concepts of extortion and deterrence generally cannot be understood except by denying the utility-comparison criterion. It is indeed the asymmetries in the threat situation, as between the two players, that make threats a rich subject for study; but the relevant asymmetries include those in the communication system, in the enforcibility of threats and of promises, in the speed of commitment, in the rationality of expected responses, and, finally (in some cases) in the relative-damage criterion.

PROMISES

Enforcible promises cannot be taken for granted. Agreements must be in enforcible terms and involve enforcible types of behavior. Enforcement depends on at least two things — some authority somewhere to punish or coerce and an ability to discern whether punishment or coercion is called for. The postwar discussions of disarmament proposals and inspection schemes indicate how difficult it may be, even if both sides should desperately desire to reach an enforcible agreement or find a persuasive means of enforcement. The problem is compounded when neither party trusts the other and each recognizes that neither trusts the other and that neither can therefore anticipate the other's compliance. Many of the technical problems of arms inspection would disappear if there were some earthly means of making enforcible promises or if the nations of the world all rendered unquestioned allegiance to some unearthly authority. But, since noncompliance may be undetectable, promises of compliance could not be enforced even if punishment could be guaranteed. The problem is doubled by the fact that punishment cannot be guaranteed, except such punishment as can unilaterally be meted out by the other party in its act of denouncing the original agreement. Furthermore, some seemingly desirable agreements must be left out for being undefinable operationally; agreements not to discriminate against each other will work only if defined in objective terms capable of objective supervision.

Promises are generally thought of as bilateral (contractual) commitments, given against a quid pro quo that is often a promise in return. But there is incentive for a unilateral promise when it provides inducement to the other player to make a choice in the mutual interest. In the left-hand matrix of Fig. 12, if choices

FIG. 12

are to be simultaneous, only a *pair* of promises can be effective; in the right-hand matrix, Row's promise brings its own reward: Column can safely choose II, yielding superior outcomes for both players. (If, in the left-hand matrix, moves are in turn, the player who chooses *second* must have the power to promise. If the players are themselves to agree on the order of moves and only one of the two can issue promises, they can agree that the other one move first. These promises, in contrast to those for the right-hand matrix, must be conditional on the second player's performance. A unilateral unconditional promise does the trick on the right-hand side but not on the left with moves in turn.) The witness to a crime has a motive for unilateral promise if the criminal would kill to keep him from squealing.[11] A nation known to be on the threshold of an absolutely potent surprise-attack weapon may have reason to foreswear it unilaterally — if there is any possible way to do so — in order to forestall a desperate last-minute attempt by an enemy to strike first while he still has a chance.

The exact definition of a promise — for example, in distinction to a threat — is not obvious. It might seem that a promise is a commitment (conditional or unconditional) that the second party welcomes, one that is mutually advantageous, as in both the games shown in Fig. 12. But Fig. 13 shows a situation in which

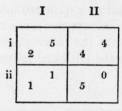

FIG. 13

Row must couple a threat and a promise; he threatens ii against I and promises i in the event of II. The promise insures Column a payoff of 4 rather than zero, once he has made a choice of II, and in that sense it is favorable to him; it does so at a cost of 1 unit to Row. But, if Row could not make the promise, Column would win 5; he would because the threat would be ineffectual without the promise, and the threat would not be incurred. A threat of ii against I by itself is no good; it cannot force Column to choose II, since a choice of II leaves him with an outcome at ii,II, zero instead of 1. Row's threat can work only if the promise goes with it; the net effect of the promise is to make the threat work, yielding Column 4 instead of 5, gaining 5 rather than 2

[11] This notion is celebrated in "Wet Saturday," by John Collier, recently reproduced by Alfred Hitchcock on TV. An inadvertent eavesdropper on a murder is ordered at gunpoint to seal his lips by leaving his own fingerprints and other incriminating evidence, so that if the body is found he will be charged with the murder. He should have insisted, however, on fabricating the evidence so as to share the guilt with the actual murderer; as it was, he got badly cheated. (*Short Stories from the "New Yorker"* [London, 1951], pp. 171–178.)

for Row. One cannot force spies, conspirators, or carriers of social diseases to reveal themselves solely by the *threat* of a relentless pursuit that spares no cost; one must also promise immunity to those that come forward.[12]

A better definition, perhaps, would make the promise a commitment that is controlled by the second party, that is, a commitment that the second party can enforce or release as he chooses. But timing is important here. The promise just discussed will work *after* the threat is fully committed; but if the victim of the promise (Column) can renounce the promise in advance, so that Row knows that Column expects zero if he chooses II, the threat itself is deterred. And, if the threat and promise are contrived in such a way as to be "legally" inseparable or if they are accomplished by some irreversible maneuver, the definition becomes obscured. (In fact, the definition breaks down whenever the equivalent of a promise is obtained by some irrevocable act rather than by a "legal" commitment.)

Actually, whenever the alternative choices are more than two, threat and promise are likely to be mixed in any "reaction pattern" that one presents to the other. So it is probably best to consider the threat and the promise to be names for different aspects of the same tactic of selective and conditional self-commitment, which in certain simple instances can be identified in terms of the second party's interest.

Enforcement schemes. Agreements are unenforcible if no outside authority exists to enforce them or if noncompliance would be inherently undetectable. The problem arises, then, of finding forms of agreement, or terms to agree on, that provide no incentive to cheat or that make noncompliance automatically visible or that incur the penalties on which the possibility of enforcement rests. While the possibility of "trust" between two partners need not be ruled out, it should also not be taken for granted; and even trust itself can usefully be studied in game-theoretic terms. Trust is often achieved simply by the continuity of the relation between parties and the recognition by each that what he might

[12] Somewhat related is the grant of immunity that strips a reticent witness of protective danger of self-incrimination, and so opens him to the ordinary sanction of contempt proceedings.

gain by cheating in a given instance is outweighed by the value of the tradition of trust that makes possible a long sequence of future agreement. By the same token, "trust" may be achieved for a single discontinuous instance, if it can be divided into a succession of increments.

There are, however, particular game situations that lend themselves to enforcible agreement. One is an agreement that depends on some kind of coordination or complementarity. If two people have disagreed on where to meet for dinner; if two criminal accomplices have disagreed on what joint alibi to give; or if members of a business firm or football team have disputed about what prices they will quote or what tactic they will follow, they nevertheless have an overriding interest in the ultimate consistency of their actions. Once agreement is formally reached, it constitutes the only possible focal point for the necessary subsequent tacit collaboration; no one has a unilateral preference now to do anything but what he is expected to do. In the absence of any other means of enforcement, then, parties might be well advised to try to find agreements that enjoy this property of interdependent expectations, even to the extent of importing into their agreement certain elements whose sole purpose is to create severe jeopardy for noncoordination. Tearing the treasure map in half or letting one partner carry the gun and the other the ammunition is a familiar example.

The institution of *hostages* is an ancient technique that deserves to be studied by game theory, as does the practice of drinking wine from the same glass or of holding gang meetings in places so public that neither side could escape if it subjected the other to a massacre. The reported use of only drug addicts as agents or employees in a narcotics ring is a fairly straightforward example of a unilateral hostage.

Perhaps a sufficient interchange of populations between nations that hate each other or an agreement to move the governing agencies of both countries to a single island where they would occupy alternate blocks of the city could be resorted to if both sides became sufficiently desperate to avoid mutual destruction. A principal drawback to the exchange of hostages, on the assumption of rational behavior, is the inherent unknowability of each other's value system adverted to earlier. The king who sends his daughter as a hostage to his enemy's court may be incapable of assuaging his enemy's fears that he really dislikes the girl. We could probably guarantee the Russians against an American surprise attack by having the equivalent of "junior year abroad" at the kindergarten level: if every American five-year-old went to kindergarten in Russia — in American establishments constructed for the purpose, designed solely for "hostage" purposes and not for cultural interchange — and if each year's incoming group arrived before the graduating class left, there would not seem to be the slightest chance that America would ever initiate atomic destruction in Russia. We cannot be quite sure that the Russians would be quite sure of this. Nor can we be quite sure that a reciprocal program would be as much of a deterrent to the Russian government; unfortunately, even if the Russian government were bound by the fear of harming Russian children, it seems nearly impossible for it to persuade us so. Still, in many surprise-attack situations a unilateral promise is better than none; and the idea of hostages may be worth considering, even when symmetrical exchanges do not seem available.[13]

Actually, the hostage idea is logically identical with the notion that a disarmament agreement between the major powers might be more efficacious (and probably more subject to technical control) if it related to *defensive* weapons and structures. To eschew defense is, in effect, to make hostages of your entire population without bothering to put them physically into the other's possession. Thus we can put our children at the mercy of the Russians and receive similar power over Russian children not only by physically trading them, with enormous discomfort and breach of constitutional rights, but also by simply agreeing to leave them so unprotected that the other can do them as much

[13] The precise definition of hostages is a little difficult. They seem to be as pertinent to threats as to promises: the American divisions that were stationed in Europe principally to demonstrate that America could not avoid becoming engaged in a European conflict can probably be viewed as hostages; if they cannot, their wives and children can, and perhaps their wives and children have been a more persuasive commitment or "trip wire" than the troops themselves. As a general rule, invaders may have to avoid the peak tourist season in countries they covet, to avoid provoking the countries that have yielded inadvertent hostages.

damage where they are as if he had them in his grasp. Thus the "balance of terror" that is so often adverted to is — if, in fact, it exists and is stable — equivalent to a total exchange of all conceivable hostages. (The analogy requires that the balance be stable, i.e., that neither side be able, by surprise attack, to destroy the other's power to strike back, but just able to inflict a surfeit of civilian agony.) [14]

Denial of enforcement. Enforcement of promises is also relevant to the influence of a third party that wishes to make an efficient outcome more difficult for the other two players. A potent means of banning illegal activities has often been the outlawing of them, so that contracts became unenforcible. Failure to enforce gambling contracts or contracts in restraint of trade or contracts for the delivery of liquor during prohibition has always been part of the process of discouraging the activities themselves. Sometimes, of course, prohibition of this sort delivers enormous power into the hands of anyone who can enforce contracts or make enforcible promises.[15] The denial of copyright liquor labels during prohibition meant that only the bigger gangs could guarantee the quality of their liquor and hence assisted them in developing monopoly control of the business. By the same token, laws to protect brands and labels can perhaps be viewed as devices that facilitate business based on unwritten contracts.

RELINQUISHING THE INITIATIVE

What makes the threat or ordinary commitment a difficult tactic to employ and an interesting one to study is the problem of finding a means to commitment, the available "penalty" to invoke against one's own nonperformance. There is consequently a related set of tactics that consists of maneuvering one's self into a position in which one no longer has any effective choice over how he shall behave or respond. The purpose of these tactics

[14] This concept is developed at length in Chapter 10.

[15] It has been argued that an important function of the racketeer is sometimes to help enforce agreements that are beyond the law. Price-cutting in the Chicago garment trade was punishable by explosion — the fee for the explosion being paid by the price-fixing organization — according to R. L. Duffus, "The Function of the Racketeer," *New Republic* (March 27, 1929), pp. 166–68.

is to get rid of an embarrassing initiative, making the outcome depend solely on the other party's choice.

This is the kind of tactic that Secretary of State John Foster Dulles was looking for in the following passage:

> In the future it may thus be feasible to place less reliance upon deterrence of vast retaliatory power. . . . Thus, in contrast to the 1950 decade, it may be that by the 1960 decade the nations which are around the Sino-Soviet perimeter can possess an effective defense against full-scale conventional attack and thus confront any aggressor with the choice between failing or himself initiating nuclear war against the defending country. Thus the tables may be turned, in the sense that instead of those who are non-aggressive having to rely upon all-out nuclear retaliatory power for their protection, would-be aggressors would be unable to count on a successful conventional aggression, but must themselves weigh the consequence of invoking nuclear war.[16]

The distinction between the type of deterrence he imputes to the 1950's and the type he imputes to the 1960's differs in the matter of who has to make that final decision; and the difference is important because the United States cannot find, or bring itself to trust, a persuasive means of commitment to the threat of massive retaliation against certain types of aggression.

There was a time, shortly after the first atomic bomb was exploded, when there was some journalistic speculation about whether the earth's atmosphere had a limited tolerance to nuclear fission; the idea was bruited about that a mighty chain reaction might destroy the earth's atmosphere when some critical number of bombs had already been exploded. Someone proposed that, if this were true and if we could calculate with accuracy that critical level of tolerance, we might neutralize atomic weapons for all time by a deliberate program of openly and dramatically exploding $n - 1$ bombs.

This tactic of shifting responsibility to the other player was

---

[16] J. F. Dulles, "Challenge and Response in U. S. Policy," *Foreign Affairs* (October, 1957). Very similar language is used by Dean Acheson (*Power and Diplomacy* [Cambridge, Mass., 1958], pp. 87–88) in discussing the role of a sizable defense force in Europe: by requiring of the enemy a major attack, rather than a small one, it makes him believe that retaliation would ensue, because "he would be making the decision for us. . . . A defense in Europe of this magnitude will pass the decision to risk everything from the defense to the offense."

nicely accomplished by Lieutenant Colonel (then Major) Steven-
son B. Canyon, U.S.A.F., in using his aircraft to protect a
Chinese Nationalist surface vessel about to be captured by Communist surface forces in his comic strip. Unwilling and unauthor-
ized to initiate hostilities and knowing that no threat to do so
would be credited, he directed his planes to jettison gasoline in a
burning ring about the aggressor forces, leaving to them the last
clear chance of reversing their engines to avoid the flames. He
could neither drop gasoline on the enemy ships nor threaten to;
so he dropped the initiative instead.

The same tactic is involved in those dramatic forms of "passive
resistance" that might be better called "active nonresistance."
According to *The New York Times*, "Striking railway workers sat
down on the tracks at more than 300 stations in Japan today,
halting 48 passenger and 144 freight trains." [17]

A more dramatic instance, also Japanese, was reported in the
same paper: "A public debate is being held here this week on
whether to send a 'suicide sit-down fleet' to the forbidden waters
around Christmas Island, the site of the forthcoming British
hydrogen bomb experiment. . . . The first object of the expedi-
tion would be to prevent the British blast." [18]

IDENTIFICATION

An important characteristic of any game is how much each
side knows about the other's value system; but a similar information problem arises with respect to sheer identification. The
bank employee who would like to rob the bank if he could only

---

[17] "Rail Strikers Sit in Tracks," *The New York Times* (May 13, 1957), pp.
14L f. The appropriate countertactic seems to be the following: The engineer
sets the throttle for slow forward speed, conspicuously climbs down from his
cab and jumps off the moving train, walks through the station and jumps back
on his engine when it catches up with him. The weakness of his position while
he is driving the train is that he can stop it more quickly than his adversaries
can get off the tracks, particularly if they have arranged to crowd themselves so
that they could not vacate the track quickly. They can forestall his countertactic
by locking themselves to the tracks and throwing away the key — if they can
persuasively inform the engineer of this before he has relinquished his own control of the engine.

[18] "Japan Debating Atomic 'Suicide,' " *The New York Times* (March 5,
1957), p. 16.

find an outside collaborator and the bank robber who would like to rob the bank if only he could find an inside accomplice may find it difficult to collaborate because they are unable to identify each other, there being severe penalties in the event that either should declare his intentions to someone who proved not to have identical interests. The boy who is afraid to ask a girl for a date because she might rebuff him is in a similar position. Similarly, the kidnaper cannot operate properly if he cannot tell the rich from the poor in advance; and the antisegregation minority in the South may never know whether it is large or small because of the penalties on declaration.

Identification, like communication, is not necessarily reciprocal; and the act of self-identification may sometimes be reversible and sometimes not. One may achieve more identification than he bargained for, once he declares his interest in an object. A nice example occurs in Shakespeare's *Measure for Measure*. Angelo, acting in place of the Duke, has a prisoner whom he proposes to kill. He could torture him, but he has no incentive to. The victim has a sister, who arrives to plead for his life. Angelo, finding the sister attractive, proposes a dishonorable bargain; the sister declines, Angelo then threatens to torture the brother unless the sister submits. At this point the game has been expanded simply by the establishment of identity and of a line of communication. Angelo's only interest in torturing the brother is in what he may gain by making a threat to do so; once there is somebody available to whom the threat can profitably be communicated, the possibility of torture has value for Angelo — not the torture itself, but the threatening of it. The sister has gotten negative value out of her trip; having identified her interest and made herself available to receive the threatening message, she has been forced to suffer what she would not have had to suffer if she had never made her identity known or if she could have disappeared into the crowd before the threat was made.

A nice identification game was uncovered in a New York suburb a few years ago. Certain motorists carried identity cards which identified them to policemen as members in a club; if the motorist with a membership card was arrested, he simply showed the card to the policeman and paid a bribe. The role of these cards was to identify the motorist as a person who, if the bribe was received, would keep quiet. It identified the motorist as a man whose promise was enforcible. But the card identifies the motorist only *after* he has been arrested; if the police could identify card-carrying motorists by looking at them, they could concentrate their arrests on card-carrying drivers, threatening a ticket unless payment were received. The card is contingent identification, at the option of the motorist. A similar situation — pertinent to the discussion of promises as well as to identification — is described by Sutherland: "Most coppers are more or less fair in their dealings with thieves simply because it pays them to be so. They will extend favors even after a pinch which they would not extend to nonprofessionals whom they lock up. They realize that it is safe to do this and that high officials will not be informed, as might be the case if favors were extended to amateurs." [19]

Identification is also relevant to an important economic fact that tends to be ignored in the conventional economics of production and exchange, namely, the enormous potential for destruction that is available and that is relevant because of the extortionate threats that could be supported by it. The ordinary healthy high-school graduate, of slightly below average intelligence, has to work fairly hard to produce more than $3,000 or $4,000 of value per year; but he could destroy a hundred times that much if he set his mind to it, according to the writer's hasty calculations. Given an institutional arrangement in which he could generously abstain from destruction in return for a mere fraction of the value that he might have destroyed, the boy clearly has a calling as an extortionist rather than as a mechanic or clerk. It is fortunate that extortion usually depends on self-identification and overt communication by the extortionist himself.

The importance of self-identification is attested by the significance attached to the doctrine that an accused person should be permitted to know and to confront his accuser. It is also reflected in secret testimony before a Grand Jury, in cases where identifiable witnesses might be intimidated by potential defendants,

[19] E. H. Sutherland, *The Professional Thief* (Chicago, 1954), p. 126.

and in efforts to keep secret the identity of eyewitnesses to a crime until the criminal is apprehended. (The strategy of law and of law enforcement and criminal deterrence is a rich field for the application of game theory.)

DELEGATION

Another "move" that is sometimes available is the delegation of part or all of one's interest, or part or all of one's initiative for decision, to some agent who becomes (or perhaps already is) another player in the game. Insurance schemes permit the sharing of interests; the insurance company has a different incentive structure from the insured party and may be better able to make threats or resist them for that reason. Requiring several signatures on a check accomplishes a similar purpose. The use of a professional collecting agency by a business firm for the collection of debts is a means of achieving unilateral rather than bilateral communication with its debtors and of being therefore unavailable to hear pleas or threats from the debtors. Providing ammunition to South Korean troops or giving them access to prisoner-of-war camps so that they can unilaterally release prisoners is a tactical means of relinquishing an embarrassing power of decision — embarrassing because it subjects one to coercive or deterrent threats or leaves one the capacity to back out of his own threat, hence the incapacity to make the threat persuasive.

The mutual-defense agreement with the Nationalist government of China is probably to be viewed partly as a means of shifting the decision for response to someone whose resolution would be less doubtful; and more recently the proposal to put nuclear weapons in the hands of European governments has been explicitly argued on grounds that it would enhance deterrence by giving the visible power to retaliate to countries that might in certain contingencies be thought less irresolute than the United States.

The use of thugs and sadists for the collection of extortion or the guarding of prisoners, or the conspicuous delegation of authority to a military commander of known motivation, exemplifies a common means of making credible a response pattern that the original source of decision might have been thought to shrink from or to find profitless, once the threat had failed. (Just as it would be rational for a rational player to destroy his own rationality in certain game situations, either to deter a threat that might be made against him and that would be premised on his rationality or to make credible a threat that he could not otherwise commit himself to, it may also be rational for a player to select irrational partners or agents.)

In the matrix in Fig. 14 — disregarding the numbers in parentheses — if Row has second move, he loses in the lower right-hand

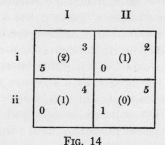

FIG. 14

corner, Column gaining his own preferred outcome. If a third party without power of decision is scheduled to receive, as a by-product, the payoff in parentheses, Row can win if some means is available for irreversibly surrendering his move to the third player. The payoffs of the latter are such that with second move he wins in the upper left-hand corner, leaving the original Row-player a payoff of 5 as a by-product. (If the third party's rewards had to be financed by Row, whose own payoffs were correspondingly reduced, it would still be worth his while to make an irrevocable assignment of portions of his various payoffs to the third player, together with assignment of the decision; with the figures shown, he would still carry away a net value of 3 in the upper left-hand corner, in contrast to 1 in the lower right.)

MEDIATION

The role of mediator is another element for analysis in game theory. A mediator, whether imposed on the game by its original

rules or adopted by the players to facilitate an efficient outcome, is probably best viewed as an element in the communication arrangements or as a third player with a payoff structure
of his own who is given an influential role through his control
over communication. But a mediator can do more than simply
constrain communications — putting limits on the order of
offers, counter-offers, and so forth — since he can invent contextual material of his own and make potent suggestions. That is,
he can influence the other player's expectations on his own initiative, in a manner that both parties cannot help mutually recognizing. When there is no apparent focal point for agreement, he
can create one by his power to make a dramatic suggestion. The
bystander who jumps into an intersection and begins to direct
traffic at an impromptu traffic jam is conceded the power to
discriminate among cars by being able to offer a sufficient increase in efficiency to benefit even the cars most discriminated
against; his directions have only the power of suggestion, but
coordination requires the common acceptance of some source of
suggestion. Similarly, the participants of a square dance may all
be thoroughly dissatisfied with the particular dances being called,
but as long as the caller has the microphone, nobody can dance
anything else. The white line down the center of the road is a
mediator, and very likely it can err substantially toward one side
or the other before the disadvantaged side finds advantage in
denying its authority. The principle is beautifully illustrated by
the daylight-saving-time controversy; a majority that wants to
do everything an hour earlier just cannot organize to do it unless it gets legislative control of the clock. And when it does, a
well-organized minority that opposed the change is usually quite
unable to offset the change in clock time by any organized effort
to change the nominal hour at which it gets up, eats, and does
business.

Mediators can also be a means by which rational players can
put aside some of their rational faculties. A mediator can con-
summate certain communications while blocking off certain facilities for memory. (In this regard he serves a function that can be
reproduced by a computing machine.) He can, for example, compare two parties' offers to each other, declaring whether or not the

offers are compatible without revealing the actual offers. He is a scanning device that can suppress part of the information put into it. He makes possible certain limited comparisons that are beyond the mental powers of the participants, since no player can persuasively commit himself to forget something.

The problem of persuasively denying one's self the knowledge that one receives by the left hand, while actively seeking it with the right hand, is nicely illustrated by the efforts of parts of governments to obtain accurate data on incomes for the purpose of statistical programs, while another part of the government is seeking the same data in order to impose taxes or to prosecute evasion. Governments have found it important to seek ways of guaranteeing that the statistical agency will deny the information it receives to the taxing agency, in order to receive the information in the first place. An analogous case of relying on an explicit mediator is that of companies that turn trade secrets over to a statistical bureau that is committed to destroy the individual data after computing the sums and averages that it will make public for the benefit of the contributing companies, or of public opinion services that suppress potentially embarrassing individual data on political or sexual practices, publishing only the aggregates. The use of mediators to forestall identification seems to be a common tactic when a buyer of large resources thinks a painting or a right-of-way can be bought cheap if the owner is unaware who it is that is interested.

Mediators may be converted into arbitrators by the irrevocable surrender of authority to him by the players. But arbitration agreements have to be made enforcible by the players' deliberately incurring jeopardy, providing the referee with the power to punish or surrendering to him something complementary to their own value systems. In turn, they must be able to trust him or to extract an enforcible promise from him. But in any case he increases the totality of means for enforcing promises: two people who do not trust each other may find a third person that they both trust, and let him hold the stakes.[20]

---

[20] I have been told that in countries where no strong tradition of business morality exists, a few partners or directors for a business may deliberately be chosen from another culture where simple honesty and fairness are considered

COMMUNICATION AND ITS DESTRUCTION

Many interesting game tactics and game situations depend on the structure of communication, particularly asymmetries in communication and unilateral options to initiate communication or to destroy it. Threats are no good if they cannot be communicated to the persons for whom they are intended; extortion requires a means of conveying the alternatives to the intended victim. Even the threat, "Stop crying or I'll give you something to cry about," is ineffectual if the child is already crying too loud to hear it. (It sometimes appears that children know this.) A witness cannot be intimidated into giving false testimony if he is in custody that prevents his getting instructions on what to say, even though he might infer the sanction of the threat itself.

When the outcome depends on coordination, the timely destruction of communication may be a winning tactic. When a man and his wife are arguing by telephone over where to meet for dinner, the argument is won by the wife if she simply announces where she is going and hangs up. And the status quo is often preserved by a person who evades discussion of alternatives, even to the extent of simply turning off his hearing aid.

As discussed in the earlier part of this chapter, mob action often depends on communication in a way that makes it possible for the authorities to obstruct mob action by forbidding groups of three or more to congregate. But mobs can themselves intimidate the authorities if they are able to identify them and to communicate with them. Even a tacit threat of subsequent ostracism or violence may be communicated from a riotous mob to the local police, if the police are known to them and are persons who have to reside among them when the occasion is over. In that case the use of outsiders may forestall the mob's intimidating threats against the authorities, partly by reducing the subsequent occasion for carrying out the threat but partly also through the difficulty of tacit communication between mob and police. Federal troops in Little Rock may have enjoyed some immunity to intimidation just by being outside the tacit communication struc-

---

to be common traits or where a reputation for them is considered of much higher value.

ture of the local populace and being patently less conversant with the local value system than were the local police. State troops were dramatically successful in quelling the Detroit race riot of 1943, when the local police were ineffectual. The use of Moors, Sikhs, and other foreign-language troops against local uprising may owe some of its success to their poor capacity to receive the threats and promises that the enemies or victims might otherwise seek to convey. Even the isolation of officers from enlisted men in military service may tend to make officers less capable of receiving and perceiving threats, hence less capable of being effectively threatened, and thus deterring intimidating threats themselves.

It is important, of course, whether or not the threatener knows that his threat cannot be received; for if he thinks it can, and it cannot, he may make the threat and fail in his objective, being obliged to carry out his threat to the subsequent disadvantage of both himself and the one threatened. So the soldiers in quelling the riot should not only be strangers and not only keep moving sufficiently to avoid "acquaintance" with particular portions of the mob; they should behave with an impassivity to demonstrate that no messages are getting through. They must catch no one's eye; they must not blush at the jeers; they must act as if they cannot tell one rioter from another, even if one has been making himself conspicuous. Figuratively, if not literally, they should wear masks; even the uniform contributes to the suppression of identification and so itself makes reciprocal communication difficult.

Conveyance of evidence. "Communication" refers to more than the transmission of messages. To communicate a threat, one has to communicate the commitment that goes with it, and similarly with a promise; and to communicate a commitment requires more than communication of words. One has to communicate *evidence* that the commitment exists; this may mean that one can communicate a threat only if he can make the other person see something with his own eyes or if he can find a device to authenticate certain allegations. One can send a signed check by mail, but one cannot demonstrate over the telephone that a check bears an authentic signature; one may show that he has a loaded gun but not prove it by simply saying so. From a game-theory point of view, the Paris *pneumatique* differs from an ordinary telegraph system, and television differs from radio. (One role of a mediator may be to authenticate the statements that the players make to each other; for example, a code system for identification might make it possible for people to transmit funds orally by telephone, the recipient being assured by the bank's code response that it is in fact the bank at the other end of the line assuring him that the payer has been identified by code and that the transaction is complete.) The importance and the difficulty of communicating evidence is exemplified by President Eisenhower's "open-skies" proposal and other suggested devices for dealing with the instability that may be caused by the reciprocal fear of surprise attack. Leo Szilard has even pointed to the paradox that one might wish to confer immunity on foreign spies rather than subject them to prosecution, since they may be the only means by which the enemy can obtain persuasive evidence of the important truth that we are making no preparations for embarking on a surprise attack.[21]

It is interesting to observe that political democracy itself depends on a game structure in which the communication of evidence is impossible. What is the secret ballot but a device to rob the voter of his power to sell his vote? It is not alone the secrecy, but the *mandatory* secrecy, that robs him of his power. He not only *may* vote in secret, but he *must* if the system is to work. He must be denied any means of proving which way he voted. And what he is robbed of is not just an asset that he might sell; he is stripped of his power to be intimidated. He is made impotent to meet the demands of blackmail. There may be no limit to violence that he can be threatened with if he is truly free to bargain away his vote, since the threatened violence is not carried out anyway if it is frightening enough to persuade him. But when the voter is powerless to prove that he complied with the threat, both he and those who would threaten him know that any punishment would be unrelated to the way he actually voted. And the threat, being useless, goes idle.

[21] L. Szilard, "Disarmament and the Problem of Peace," *Bulletin of the Atomic Scientists*, 2:297–307 (October, 1955).

An interesting case of tacit and asymmetrical communication is that of a motorist in a busy intersection who knows that a policeman is directing traffic. If the motorist sees, and evidently sees, the policeman's directions and ignores them, he is insubordinate; and the policeman has both an incentive and an obligation to give the man a ticket. If the motorist avoids looking at the policeman, cannot see the directions, and ignores the directions that he does not see, taking a right of way that he does not deserve, he may be considered only stupid by the policeman, who has little incentive and no obligation to give the man a ticket. Alternatively, if it is evident that the driver knew what the instructions were and disobeyed them, it is to the policeman's advantage not to have seen the driver, otherwise he is obliged, for the reputation of the corps, to abandon his pressing business and hail the driver down to give him a ticket. Children are skilled at avoiding the receipt of a warning glance from a parent, knowing that if they perceive it the parent is obliged to punish noncompliance; adults are equally skilled at not requesting the permission they suspect would be denied, knowing that explicit denial is a sterner sanction, obliging the denying authorities to take cognizance of the transgression.[22]

The efficacy of the communication structure can depend on the kinds of rationality that are imputed to the players. This is illustrated by the game situation known as "having a bear by the tail." The minimum requirement for an efficient outcome is that

[22] What might be called the "legal status" of communication is nicely developed by Goffman: "Tact in regard to face-work often relies for its operation on a tacit agreement to do business through the language of hint — the language of innuendo, ambiguities, well placed pauses, carefully worded jokes, and so on. The rule regarding this unofficial kind of communication is that the sender ought not to act as if he had officially conveyed the message he has hinted at, while the recipients have the right and the obligation to act as if they have not officially received the message contained in the hint. Hinted communication, then, is deniable communication." He refers to the "unratified" participation that can occur in spoken interaction: "A person may overhear others unbeknown to them; he can overhear them when they know this to be the case and when they choose either to act as if he were not overhearing them or to signal to him informally that they know he is overhearing them." He points out that the obligation to respond, for example, to an insulting remark that one has inadvertently overheard may depend on whether the overhearing has acquired "ratification" (pp. 224, 226).

the bear be able to incur an enforcible promise and that he be able to transmit credible evidence that he is committed, either by a penalty incurred or by a maneuver that destroys his power not to comply (like extracting his own teeth and claws). But if the bear is of limited rationality, having a capacity for making rational and consistent choices among the alternatives that he perceives but lacking the capacity to solve games — that is, lacking the capacity to determine introspectively the choices that a partner would make — the communication system must make it possible for him to receive a message from his partner. The partner must then formulate the proposition (choice) for the bear and communicate it to him, in order that the bear may then respond by accepting the promise (now that he sees what the "solution" is) and transmitting authoritative evidence back to his own partner.

INCORPORATION OF MOVES IN A GAME MATRIX

One is led to suppose that, if a game has potential moves like threats, commitments, and promises that are susceptible of formal analysis, it must be possible to represent such moves in the traditional form of strategy choices, with the payoff matrix of the original game expanded to allow for the choices among these various moves.

The first point to observe is that a commitment, a promise, or a threat can usually be characterized in a fashion equivalent to the following: to make one of these moves, a player selectively reduces — visibly and irreversibly — some of *his own* payoffs in the matrix. This is what the move amounts to.[23] We could also say that one openly selects a strategy in advance for responding to the other's choice; but more than selection is required. The player must invoke penalty on his own failure to pursue subsequently the particular strategy of response that he has selected beforehand. And to invoke a penalty on failure to follow a strategy is mathematically equivalent to subtracting the amount of

---

[23] Daniel Ellsberg, some of whose work in the field of strategy was contained in the lectures mentioned in Chapter 1, independently arrived at precisely this formulation of the threat or commitment, namely, as a selective reduction of some of one's own payoffs in the strategy matrix.

the penalty from one's own payoffs in all cells that do not correspond to the strategy so selected.[24]

Specifically, in Fig. 15 *A,* Row would commit himself to ii by subtracting from his own payoffs in the first row sufficiently large quantities — 5 in the example shown — to make ii a domi-

	A			B			C	
	I	II		I	II		I	II
i	5 / 2	0 / 1	i	5 / −3	0 / −4	i	5 / 2	0 / 1
ii	1 / 0	2 / 5	ii	1 / 0	2 / 5	ii	1 / −5	2 / 0

FIG. 15

nant strategy, that is, a strategy that he would follow no matter which column the other player selects. The result would be the modified matrix shown in Fig. 15 *B.* (Committing himself to i

[24] Threats, promises, and unconditional commitments have already been illustrated; a more general "reaction function" is illustrated in the accompanying matrix. If Row can attach adequate penalties to his own selection of any cells other than those starred, he leaves Column a simple maximization problem which Column solves by choosing his third strategy. Row has "won" almost his favorite cell; specifically, he has secured for himself the most favorable cell among those that leave Column no lower than his "minimax" value. This is the generalization of the tactic that, for simple two-way or three-way choices, can be identified as a "commitment," "threat," "promise," or combination of them. (Further generalization would include randomized strategies; these are introduced in Chapter 7.)

	I	II	III	IV	V
i	6 / 1	10 / 11	2 / 10	9 / 2	7 / 10
ii	9 / 8	4 * / 12	0 / 25	1 / 20	15 / 3
iii	20 / 9	15 / 2	6 * / 16	1 * / 18	17 / 14
iv	2 * / 6	10 / 8	7 / 7	4 / 5	3 * / 20

with penalty of 5 would yield the matrix in Fig. 15 *C*.) Can we now build up a larger matrix that represents not only the actual choices of rows and columns in the original game, such as those in Fig. 15 *A*, but also the strategies of *commit, threaten, promise,* and so forth? Certainly, once we have specified what moves are available and the order in which they are to be taken. Take the simple game in which Row has the power to commit himself visibly in advance, and Column has first move in the *original* game, that is, chooses his column before Row makes his *final* choice of row.

Originally Row, having second move, had four strategies available. He could pick i no matter what; he could pick ii no matter what; he could play i to column I and ii to column II; or he could play ii to column I and i to column II. Including the possibility of commitment, he now has *first* the choice of committing himself; and to each of these first choices he can attach any one of the four strategies just mentioned for his final move. For example, he can commit himself to ii and play ii no matter what; he can commit himself to ii and play i no matter what; he can commit himself to ii and play i to column I, ii to column II; or he can commit himself to ii and play ii to column I, i to column II. Altogether, he has twelve possible strategy combinations.

Column has eight possible strategy combinations: for each of three contingencies he has either of two moves, the moves being I and II, the contingencies being Row's commitment to i, Row's commitment to ii, and Row's noncommitment.

If we put these strategies into matrix form, we get Fig. 16. The 12 × 8 matrix of Fig. 16 represents the tacit ("noncooperative") game that corresponds to the players' private decisions on *how to play* the original game. The eight possible strategies available to Column, for example, can be thought of as the eight possible distinct *sets of complete instructions* that he might give an agent who would then play the original game for him — that is, play the game at which he chooses one of two columns, depending on whether and how Row committed himself first. There is no loss to either player in being supposed to play this enlarged game tacitly, since what would have been each player's *adaptations* to the other's prior moves is now fully allowed for in the specifica-

		I	II	III	IV	V	VI	VII	VIII
		0-I 1-I 2-I	0- I 1- I 2-II	0- I 1-II 2- I	0- I 1-II 2-II	0-II 1- I 2- I	0-II 1- I 2-II	0-II 1-II 2- I	0-II 1-II 2-II
i	0; I-i, II-i	5 / 2	5 / 2	5 / 2	5 / 2	0 / 1	0 / 1	0 / 1	0 / 1
ii	0; I-ii, II-ii	1 / 0	1 / 0	1 / 0	1 / 0	2 / 5	2 / 5	2 / 5	2 / 5
iii	0; I-i; II-ii	5 / 2	5 / 2	5 / 2	5 / 2	2 / 5	2 / 5	2 / 5	2 / 5
iv	0; I-ii, II-i	1 / 0	1 / 0	1 / 0	1 / 0	0 / 1	0 / 1	0 / 1	0 / 1
v	1; I-i, II-i	5 / 2	5 / 2	0 / 1	0 / 1	5 / 2	5 / 2	0 / 1	0 / 1
vi	1; I-ii, II-ii	1 / -5	1 / -5	2 / 0	2 / 0	1 / -5	1 / -5	2 / 0	2 / 0
vii	1; I-i, II-ii	5 / 2	5 / 2	2 / 0	2 / 0	5 / 2	5 / 2	2 / 0	2 / 0
viii	1; I-ii, II-i	1 / -5	1 / -5	0 / 1	0 / 1	1 / -5	1 / -5	0 / 1	0 / 1
ix	2; I-i, II-i	5 / -3	0 / -4	5 / -3	0 / -4	5 / -3	0 / -4	5 / -3	0 / -4
x	2; I-ii, II-ii	1 / 0	*2 / 5	1 / 0	*2 / 5	1 / 0	*2 / 5	1 / 0	*2 / 5
xi	2; I-i, II-ii	5 / -3	2 / 5	5 / -3	2 / 5	5 / -3	2 / 5	5 / -3	2 / 5
xii	2; I-ii, II-i	1 / 0	0 / -4	1 / 0	0 / -4	1 / 0	0 / -4	1 / 0	0 / -4

FIG. 16

tion of strategies in the enlarged version of the game; they *are* strategies of response or adaptation.

This is brought out in the labeling of Fig. 16. As before, Column's choices in the original two-move game are labeled I and II; Row's choices, i and ii. Additionally, the symbol "2" will denote Row's commitment to row ii, "1" a commitment to row i, and "o" a decision not to commit himself. In the enlarged game, a single "strategy" for Column is now denoted by three pairs of symbols, such as o-i, i-II, 2-I, which would mean, "Choose column I if he does not commit himself, column II if he commits himself to row i, and column I if he commits himself to row 2." For Row, a strategy consists of a decision on o, i, or 2, plus a pair of symbols denoting how he will react to each of Column's possible choices. For example, i; I-i, II-i would mean, "Commit to row i, then choose row i no matter what Column does." Knowing the payoffs in the original game, Fig. 15 A, the players can identify the payoffs in the enlarged game of Fig. 16. We can imagine Row and Column, instead of meeting to play the original game, sending their agents to play for them, each agent fully instructed for all contingencies (that is, given one particular strategy for the enlarged game). To determine what instructions to give, Row and Column consider the matrix in Fig. 16; in effect, they play the tacit game in that matrix, leaving to their agents just the role of messenger.

What is the "solution" of this enlarged tacit game? Or, rather, can we identify an evident solution to the original game? And, if so, how does it show up in the enlarged matrix? The original game clearly has a solution for rational players. (A) If Row is committed to row i, with a penalty of 5 for breaking his commitment, Column can see that row i will be chosen, no matter which column he chooses; Column chooses his preferred cell in the upper row, which is the upper left cell, i,I. And Row knows that, if he commits himself to row i, he gets the payoff in that upper-left cell, which is 2. (B) If, instead, Row commits himself to row ii (subtracts 5 from his payoff in row i), Column chooses II in preference to I; and Row knows he will get 5. Finally, (C) if Row remains uncommitted, Column knows that Row will pick the highest row payoff in the column chosen; thus if Column chooses I, Row takes i, and Column gets 5; if Column takes II, Row takes ii, and Column gets 2. Column prefers I; this leaves Row a payoff of 2; and Row can anticipate it. So Row's best outcome is to commit himself to row ii. This is the evident "solution"; it has a payoff of [5 2], and it corresponds to the strategy 2; I-ii, II-ii for Row, and to all four strategies containing 2-II for Column. (What Column would have done in contingencies o and 1 is of no material consequence, once Row has made his first move.) These are the starred cells in Fig. 16, row x. (In effect, Row's first move is a choice of which to play among the three different two-move games, *A, B,* and *C,* shown in Fig. 15, in which he has second move.)

How do we characterize the cells, or pairs of strategies, that represent the "solution" in Fig. 16? They constitute a solution of the kind that has been called a *solution in the complete weak sense.*[25] It can be arrived at, within the framework of the enlarged matrix, by a process of discarding "dominated" rows and strategies. A row is dominated by another row if every payoff to Row in the dominating row is at least as good as the corresponding payoff in the dominated row and at least one payoff is better. Applying this criterion, the first row is dominated by the third, and we strike it out. (The argument might be that Row can safely eliminate the strategy represented in the first row, since the third is at least as good in every contingency and better in some.) So is the second, so is the fourth; so are all the rest except the tenth. Neither the third nor the tenth row dominates the other, so for the moment we keep them both. Comparing columns, no single column dominates another; but, having eliminated all rows but the third and tenth (arguing, perhaps, that Row would not choose them anyway), Column can make his comparison between only the third and tenth cells in the columns. Now it is apparent that the second column dominates the first, the third, the fifth, and the seventh. After striking out those columns that are dominated in the reduced set of rows, we can look again at rows iii and x. Originally, neither dominated the other; but, with the first, third, fifth, and seventh *columns* gone, the tenth row dominates the third. Striking out the third row, we are left with a single row, row x, intersected by four columns. The payoffs are the same in the four intersections, indicating that it is inconsequential which of those four strategies Column plays, as long as Row plays the tenth row. (That is, once Row has committed himself to the second row of the original 2 × 2 matrix, Fig. 15 *A,* as Column can expect him to do, it makes no differ-

[25] Compare Luce and Raiffa, pp. 106–09.

ence what instructions Column gives his agent regarding the two contingencies that did not arise.) [26]

This, then, is the way that a solution to the original *sequential-move game* shows up in the static ("moveless," or simultaneous-tacit-choice) game. It is a solution arrived at by discarding dominated strategies, with the criterion for domination reflecting only the undiscarded strategies at each stage. This seems to be the general form of solution in the enlarged tacit game that corresponds to a sequential-move game when the latter has a determinate solution. The discarding of rows and columns can actually be identified with the process of first calculating the rational *last move* for all possible sets of prior moves, then, knowing what last move would follow each next-to-last move, calculating the best next-to-last move for all possible sets of prior moves and so on back to the best first move of the game.

While it is instructive and intellectually satisfying to see how such tactics as threats, commitments, and promises can be absorbed into an enlarged, abstract "supergame" (game in "normal form"), it should be emphasized that we cannot learn anything about those tactics by studying games that are already in normal form. The objects of our study, namely, these tactics together with the communication and enforcements structures that they depend on, and the timing of moves, have all disappeared by the time the game is in normal form. What we want is a theory that systematizes the study of the various universal ingredients

---

[26] It is worth noting that the order in which we discard the rows and columns that are eligible for discard can affect the form of the "solution." In the procedure outlined in the text, we first discarded all rows but the third and tenth; we then observed that columns I, III, V, and VII, were eligible for discard, and discarded them; at that stage, row iii was seen to be dominated, and it was discarded; and we were left with row x intersected by four columns that yielded identical payoffs in that row. But we might have noted, as we discarded the four columns, that two more columns could also be discarded at that stage, namely columns VI and VIII, which show inferior payoffs to Column, *in row iii,* than columns II and IV. In other words, at that point in the process, row iii and columns VI and VIII were all eligible for discard; but if we arbitrarily choose first to eliminate row iii and *then* proceed to the columns, the two columns in question are no longer dominated. Thus, in a sense, the contents of our "solution" depend on an arbitrary choice of procedure; whether we are left with two cells with identical payoffs, however, or four cells with identical payoffs, depends on that arbitrary choice. The payoffs, however, are the same in either case. The rationale might be that at some stage

that make up the move-structure of games; too abstract a model will miss them.[27]

The matrix representation of a sequential game does help emphasize, however, that the formal "determinateness" of games that are resolved by tactical moves does not detract from their essential game-of-strategy character. A threat "wins" and determines an outcome only because it induces the other player to choose in one's favor. The other player retains his original freedom of choice; and his choice still depends on his anticipation of the threatener's final choice. The threatener's first choice — to threaten or not — thus depends on what he expects the threat-

---

Column sees that he needn't reason any further, that Row has a clearly determined choice that makes it inconsequential whether Column further narrows his decision, but that the exact point at which he perceives this, and what columns are left uneliminated when he does perceive it, depends to some extent on which of several alternative routes he pursues in his reasoning process. (If there were communication costs in narrowing his choice of strategy, Column might prefer to choose strategy 2–II only, leaving unspecified what choice would correspond to Row's strategy o or 1. If, to take a contrary case, there are risks that Row's strategy will be erroneously recorded or communicated, or unintelligently chosen, Column reduces his risks by specifying o–I as well. In the latter case he, in effect, treats row iii as not wholly unlikely in spite of its domination by row x. And if, to take the matter further, he suspects that the referee has a tendency to hear "row v" when other rows are actually chosen, he may further narrow his choice to o–I, 1–I, 2–II, the "solution" being the intersection of row x and Column II, since the intersection of v and IV is inferior to that of v and II and gives him grounds for this further refinement of his choice. In general, by attaching risks of error of various sort, or differential costs of different ways to specify a strategy, a rather richer problem is formed, and one that can lead to different conclusions. The problems treated in Chapters 7 and 9, involving certain forms of random behavior, error, or misinformation, can produce this kind of result.)

[27] Incidentally, casting a particular game into supergame matrix form is generally not a feasible technique of analysis; the number of rows and columns (that is, the number of sequential-move strategies) becomes astronomically large, even for quite simple games. To illustrate, consider a 3 × 3 matrix, with Column to choose first; add a prior opportunity for Row to commit himself to any partially or fully specified strategy of response; finally, to study the "defense" against threats, allow Column a still earlier opportunity to commit his choice of column. That is, Column may first commit himself unconditionally if he pleases, Row may then commit himself conditionally in whatever way he pleases, then Column chooses a column and finally Row chooses a row. Let us not complicate the game by limiting sizes of penalties or by inserting any uncertainty or imperfect communication system. This "simple" game, which is not terribly difficult to analyze in its extensive form, turns out to have more than a "googol" (1 followed by a hundred zeros) of columns.

ened player to expect the threatener to do. The reciprocal-expectation character of the game remains; the threat, like the unconditional commitment or like the broader concept of "reaction function" when many choices of action are available, works by constraining another player's expectations through the manipulation of one's own incentives.

THE PARADOX OF STRATEGIC ADVANTAGE

It is, of course, a corollary principle that if the payoff matrix to begin with had already shown values for one of the players reduced in the same pattern as that in which he would reduce it deliberately at the winning move, he simply wins without needing to make the move overtly. (This is the point that, in diagrammatic form, was illustrated in the final paragraph of Chapter 2, and referred to as an abstract example of the principle that, in bargaining, weakness may be strength.) There is probably no single principle of game theory that epitomizes so strikingly the mixed-motive game as this principle that a worsening of some or even all of the potential outcomes for a particular player and an improvement in none of them may be distinctly — even dramatically — advantageous for the player so disadvantaged. It explains why a sufficiently severe and certain penalty on the *payment* of blackmail can protect the potential victim, how the burning of bridges behind one's self while facing an enemy may dishearten an enemy and induce his retirement, or why a lady might, in an earlier era, defy the search party by haughtily placing the sought object in her bosom.[28]

[28] It also explains why a "promise" to abstain from a choice that would damage the other player may not be welcomed by him. A promise that *permits* him safely to make a particular choice may assure us that he *would* make it, so that we can count on it and make some prior choice that is to his disadvantage. By the same token, *adding* values selectively to the other's payoffs can absolutely worsen his position — if we have a means of making the addition. In the accompanying matrix, assuming Row has first move, Row can "win" — he can gain 7 at Column's expense — if he unilaterally guarantees to compensate Column in the event of an outcome at i,II, the compensation coming out of his own winnings. If he promises to pay 2 to Column in such an event, he gets 8; Column gets 3; otherwise, without the promised compensation, Row cannot choose i, and the outcome is at ii,I with payoffs of 1 and 10, respectively. Column obviously prefers that Row be unable to commit himself to confer the

It was reported unofficially during the Korean War that when the Treasury Department blocked Communist Chinese financial assets, it also knowingly blocked some non-Communist assets as a means of immunizing the owners against extortionate threats against their relatives still in China. Quite likely, for owners located in the United States, the very penalties on transfer of funds to Communist China enhanced their capacity to resist extortion. Deliberately putting one's own assets in a form that made evasion of the law more difficult, or lobbying for more severe penalties on illegal transfer of one's own funds, or even getting one's self temporarily identified as a Communist sympathizer so that his funds would be blocked might have been an indicated tactic for potential victims, to discourage the extortionate threat in advance.

A similar principle is reflected in Article 26 of the Japanese peace treaty, which gives the United States certain claims if subsequent Japanese territorial concessions to other powers are more favorable. When the Japanese were reported to be under pressure from the Russians for additional territorial concessions in 1956, Secretary of State John Foster Dulles pointedly described that article of the treaty in his press conference and said that he had recently "reminded the Japanese of the existence of that clause." [29] The evident intention was to strengthen Japanese resistance; and it may be supposed that by "reminding" the Russians of the same clause through the medium of his press conference, Dulles helped to provide the Japanese with the familiar bargaining claim, "If I did it for you, I'd have to do it for every-

---

"benefit." (If the blackmailer cannot scale down his demands to where what he demands, plus the fine for paying blackmail, are less than the damage he threatens, he may offer to pay his victim's fine. This guarantees what his victim's response to the threat will be; so the threat is made, to the disadvantage of the victim.)

	2	1				2	3
0		10	becomes	0		8
	10	0			10	0
1		2		1		2

[29] Transcript of the Remarks by Secretary of State Dulles at His News Conference, *The New York Times* (August 29, 1956), p. 4.

one else." It was, in terms used earlier, a "commitment" secured by the penalty of a forfeit to the United States. (Paradoxically, the United States could not give the Japanese the benefit of this bargaining gimmick unless the United States were patently motivated to take advantage of its claim if the tactic failed.) [30]

"STRATEGIC MOVES"

If the essence of a game of strategy is the dependence of each person's proper choice of action on what he expects the other to do, it may be useful to define a "strategic move" as follows: A strategic move is one that influences the other person's choice, in a manner favorable to one's self, by affecting the other person's expectations on how one's self will behave. One constrains the partner's choice by constraining one's own behavior. The object is to set up for one's self and communicate persuasively to the other player a mode of behavior (including conditional responses to the other's behavior) that leaves the other a simple maximization problem whose solution for him is the optimum for one's self, and to destroy the other's ability to do the same.

There is probably no contrast more striking, in the comparison of the mixed-motive and the pure-conflict (zero-sum) game, than the significance of having one's own strategy found out and appreciated by the opponent. Hardly anything captures the spirit of the zero-sum game quite so much as the importance of "not being found out" and of employing a mode of decision that is proof against deductive anticipation by the other player.[31] Hardly anything epitomizes strategic behavior in the mixed-motive game so much as the advantage of being able to adopt a mode of behavior that the other party will take for granted.

[30] That one's position can be painfully weakened by new legal powers is poignantly suggested by one of the arguments raised against legalizing euthanasia, granting hopeless incurables the right to authorize their own removal: "What . . . would be the effect on old people with incurable infirmities who are already suspicious that those around them want to get rid of them?" (John Beavan, "The Patient's Right to Live — and Die," *The New York Times Magazine,* August 9, 1959, pp. 14, 21–22.)

[31] Concerning this point, Von Neumann and Morgenstern say (p. 147): "We have placed considerations concerning the danger of having one's strategy found out by the opponent into an absolutely central position."

It can, of course, be an advantage in the zero-sum game to have the opponent believe firmly in a particular mode of play for one's self, but only if that belief is in error. In the mixed-motive game, one is interested in conveying the *truth* about his own behavior — if, indeed, he has succeeded in constraining his own behavior along lines that, when anticipated, win.

Another paradox of mixed-motive games is that genuine ignorance can be an advantage to a player if it is recognized and taken into account by an opponent. This paradox, which can arise either in the coordination problem or in the immunity from a threat, has no counterpart in zero-sum games. And, similarly, in a zero-sum game between rational players with full information it can never be an advantage to move first (to play the "minorant game" in the language of von Neumann and Morgenstern) ; in the mixed game it certainly can.

6

GAME THEORY AND
EXPERIMENTAL RESEARCH

The foregoing discussion suggests several conclusions about the methodology appropriate to a study of bargaining games. One is that the mathematical structure of the payoff function should not be permitted to dominate the analysis. A second one, somewhat more general, is that there is a danger in too much abstractness: we change the character of the game when we drastically alter the amount of contextual detail that it contains or when we eliminate such complicating factors as the players' uncertainties about each other's value systems. It is often contextual detail that can guide the players to the discovery of a stable or, at least, mutually nondestructive outcome. In terms of an earlier example, the ability of Holmes and Moriarty to get off at the same station may depend on the presence of something in the problem other than its formal structure. It may be something on the train or something in the station, something in their common background, or something that they hear over the loudspeaker when the train stops; and though it may be difficult to derive scientific generalizations about what it is that serves their need for coordination, we have to recognize that the *kinds* of things that determine the outcome are what a highly abstract analysis may treat as irrelevant detail.

A third conclusion, which is particularly applicable whenever the facilities for communication are short of perfect, where there is inherent uncertainty about each other's value systems or choices of strategies, and especially when an outcome must be reached by a sequence of moves or maneuvers, is that some *essential* part of the study of mixed-motive games is necessarily empirical.

This is not to say just that it is an empirical question how people do actually perform in mixed-motive games, especially games too complicated for intellectual mastery. It is a stronger statement: that the principles relevant to *successful* play, the *strategic* principles, the propositions of a *normative* theory, cannot be derived by purely analytical means from a priori considerations.

In a zero-sum game the analyst is really dealing with only a single center of consciousness, a single source of decision. True, there are two players, each with his own consciousness; but minimax strategy converts the situation into one involving two essentially unilateral decisions. No spark of recognition needs to jump between the two players; no meeting of minds is required; no hints have to be conveyed; no impressions, images, or understandings have to be compared. No social perception is involved. But in the mixed-motive game, two or more centers of consciousness are dependent on each other in an essential way. Something has to be communicated; at least some spark of recognition must pass between the players. There is generally a necessity for some social activity, however rudimentary or tacit it may be; and both players are dependent to some degree on the success of their social perception and interaction. Even two completely isolated individuals, who play with each other in absolute silence and without even knowing each other's identity, must tacitly reach some meeting of minds.

There is, consequently, no way that an analyst can reproduce the whole decision process either introspectively or by an axiomatic method. There is no way to build a model for the interaction of two or more decision units, with the behavior and expectations of those decision units being derived by purely formal deduction. An analyst can deduce the decisions of a single rational mind if he knows the criteria that govern the decisions; but he cannot infer by purely formal analysis what can pass between two centers of consciousness. It takes at least two people to test that. (Two analysts can do it, but only by using themselves as subjects in an experiment.) *Taking a hint* is fundamentally different from deciphering a formal communication or solving a mathematical problem; it involves discovering a message that has been planted within a context by someone who thinks he shares with the recipient certain impressions or associations. One cannot, without empirical evidence, deduce what understandings can be perceived in a nonzero-sum game of maneuver any more than one can prove, by purely formal deduction, that a particular joke is bound to be funny.

To illustrate, consider the question whether two people, looking at the same ink blot, can identify the same picture or suggestion in it if each is trying and knows that the other is trying to concert on the same picture or suggestion? The answer to this question can be found only by trying. But, if they can, they can do something that no *purely formal* game theory can take into account; they can do *better* than a purely deductive game theory would predict. And, if they can do better — if they can rise above the limitations of a purely formal game theory — even a normative, prescriptive, strategic theory cannot be based on purely formal analysis. We cannot build either a descriptive theory or a prescriptive theory on the assumption that there are certain intellectual processes that rational players are *not* capable of, of the kind involved in "taking a hint"; it is an empirical question whether rational players, either jointly or individually, can actually do better than a purely formal game theory predicts and should consequently ignore the strategic principles produced by such a theory.[1]

[1] A good laboratory example of the communication-perception part of game strategy is the experiment reported by M. M. Flood, who presented his players with a 2 $\times$ 2 nonzero-sum matrix for 100 consecutive tacit plays. The special property of the matrix is that the players can win only by cooperating on a particular cell on each play, but to distribute the winnings for the 100-play sequence they must cooperate on some pattern of alternation among two or more cells that discriminate differently between the two players. And the only means of negotiating over the distribution to be sought and concerting on a pattern of alternating play that achieves it is through the choices they actually make as the play proceeds. This "communication" stage — and any later stage when one player may depart from the tacitly agreed pattern to cheat a little and have to be punished by a reprisal pattern — is jointly expensive to them, since an uncoordinated choice is a lost chance to make some money. M. M. Flood, "Some Experimental Games," *Management Science*, 5:5–26 (October, 1958).

The question of how to communicate a proposal effectively and how to interpret the other player's proposal implicit in his pattern of play is evidently dependent on some mutual perception of a shared sense of pattern — a jointly recognized ability to complete a pattern of which a fragment has been displayed — not unlike the process involved in the experiments of the Gestalt psycholo-

Again it should be emphasized that the reason why this kind of consideration does not arise in the zero-sum game is that any such social interaction could not be to the advantage of both players simultaneously and that at least one of the rational players would have both motive and ability to destroy all social communication. But in a nonzero-sum game that involves any initial uncertainty over which among the possible outcomes are in fact efficient and any need for coordinated mutual accommodation to get to an efficient outcome, a rational player cannot absent himself in self-defense from the social process; he cannot turn off his hearing aid to avoid being constrained by what he hears, if complete radio silence makes efficient collaboration impossible. Nor can he rationally fail to open a letter, once it is delivered, since the other party will have assumed that he will open it and have acted accordingly.

At this point a question arises whether the game-theory trail ramifies indefinitely over the whole domain of social psychology or leads into a more limited area particularly congenial to game theory. Are there some general propositions about cooperative behavior in mixed-motive games that can be discovered by experiment or observation and that yield a widely applicable insight into the universe of bargaining situations? Although success is not assured, there are certainly some promising areas for research; and even if we cannot discover general propositions, we may at least disprove empirically some that are widely held. It does appear that game theory is badly underdeveloped from the experimental side.

Consider a game like the one described earlier, involving the movement of counters over a map, or the modified chess game

---

gists mentioned in an earlier footnote. And, while a purely formal theory of communication may derive certain minimum standards of "efficiency" in communication that rational players ought to achieve, it is an empirical question whether players can do better than that. How well one can take a hint and what kinds of hints are most successful are empirical questions of social perception, probably amenable to experimental study. (The same problem arises if two men at an auction recognize that they are jointly losing money by bidding against each other and try, without giving any overt evidence of collusion, to concert on some pattern of reciprocal and alternating abstention from bidding that both saves them money jointly and distributes the savings and the opportunities between them.)

that was made nonzero-sum. These can be taken to represent games in "limited war"; both players can gain by successfully avoiding mutually destructive strategies. Here is a game in which the ability of the two players to avoid mutual destruction may well depend on what *means* for successful coordination of intentions are provided by the incidental details of the game, by such things as a configuration of the map or board, the suggested names of the pieces, the tradition or precedent that goes with the game, and the scenario or connotative background that is instilled into the players before the game begins. It is a sufficiently complicated game to require perceptive play by both sides and the successful conveyance of intentions. If we suppose for a moment that the technical problem of constructing a playable game of that type has been mastered, it is worth while to consider what line of questions we might try to investigate or what hypotheses we might test.

One such question would be this: by and large, does it appear that the players are any more successful in reaching an efficient solution, that is, a mutually nondestructive solution, when (*a*) full or nearly full communication is allowed, (*b*) no communication or virtually none is allowed, other than what can be conveyed by the moves themselves, or (*c*) communication is asymmetrical, with one party more able to send messages than he is to receive them? There is no guaranty that a single, universally applicable answer would emerge; nevertheless, some quite general valid propositions about the role of communication might well be discovered. The enormous significance of this question is attested by some of the current controversies about whether the possibility of keeping war limited is greater if there is good communication between both sides, or if there are unilateral declarations ahead of time by one side or the other, or if there is virtually no overt communication between the belligerents.[2]

---

[2] To preclude any possible misunderstanding: the writer is not suggesting that limited war can be simulated in the laboratory or that experimental results regarding the limiting process can be directly transferred to the outside world. Experiments of the kind described would come under the heading of "basic research." And it would be concerned mainly with the perceptual and communicative side of the problem, not the motivational — except to the extent that motivations affect social perception. The probability that the results

Another set of questions, also pertinent to problems of limited war, international or other, would be whether a stable, efficient outcome is more likely when the connotations of the game — the names and interpretations that are overtly attached to the moves and pieces and objects on the board — are familiar and recognizable or when they are quite novel, unfamiliar, and unlikely to inspire similar notions in the two players. Is it — to speak of the game in a particular extensive form — more likely that rational players can keep a war limited in Southeast Asia, using conventional and atomic weapons, or in a battle against an unknown adversary on the surface of the moon, using strange bacterial weapons? These are important questions; they are at the very center of game theory; and they are questions that cannot possibly be given a confident answer without empirical evidence. And there is no arguing that rational players have the intellectual capacity to rise above these details of the game and ignore them; the importance of the details is that they can be supremely helpful to both players and that rational players know that they may be dependent on using these details as props in the course of their mutual accommodation.

Is a stable, efficient outcome more likely between two players of similar temperament and cultural background or between two quite different players? Is a stable, efficient solution more likely with two practiced players, two novices, or one novice and a practiced player; and in the latter pair, who has the advantage?

In a game of this sort, how crucial are the opening moves? If stable patterns of behavior, that is, "rules of the game," are not discovered early, will they be discovered at all? Is mutually successful play more likely if the general philosophy of each player is to begin with "tight" rules or highly "limited" weapons and resources, loosening them a little only as the occasion demands it, or if each player sets himself wider limits at the outset in order to avoid having to establish a practice of loosening rules as he goes?

---

of such research would find ready application, however, is enhanced by the observation that much current theorizing on, for instance, the role of communication in limited war or the types of limitations most likely to be observed seems itself to be based only on what might be described as implicit experimental games played introspectively.

How much influence on a game of this sort can a "mediator" have, and what kinds of mediating roles are most effective? Does it help or hinder the other two players if the mediator has a stake of his own in the outcome? To what extent can a mediator discriminate in favor of one of the two players and still increase the likelihood of a stable, efficient outcome?

It would be interesting in a game of this sort to have the players score both themselves and their partners from time to time on such matters as who is playing the more aggressively or the more cooperatively, and what "rules" each thinks are in force and thinks the other thinks are in force; of who is "winning" in a bilateral sense (it being recalled that the substantial ignorance of each other's value system makes this always a matter of interpretation); of when the game has reached a "critical" turning point, or when an "innovation" in tactics has been introduced, or when a particular move by the other side is to be interpreted as "retaliation" or a new initiative.

Because a "law of reprisal" is essentially *casuistic* in nature; because the mutually recognized restraints in any form of "limited war" are essentially based on something psychologically and sociologically akin to *tradition*; and because the received body of casuistry and tradition is often wholly inadequate to the game at hand (say, graduated atomic reprisal on the U.S.S.R. and America while limited atomic war obtains in Europe, or the bombing of grammar schools in an area without recent experience in racial violence, or the introduction of new forms of nonprice competition in a particular industry), it seems likely that the empirical part of game theory will include experimental work like that of Muzafer Sherif. He finds that when no norms exist for a laboratory judgment, they are created by the subjects; and when norms are created for two parties in the same process, each player's developing norm influences the other's. There is a process of genuine learning with respect to *values*; each side adapts its own system of values to the other's, in forming its own. When the supply of available "objective" criteria is incapable of yielding a complete set of rules, that is, when the game is "indeterminate," norms of some sort must be developed, mutually perceived, and accepted; patterns of action and response have to be legitimized.[3] In an almost unconsciously cooperative way, adversaries must reach a mutually recognized definition of what constitutes an innovation, a challenging or assertive move, or a cooperative gesture, and they must develop some common norm regarding the kind of retaliation that fits the crime when a breach of the rules occurs.[4]

A "scenario" might, for example, identify one of the players as "aggressor"; it might give the outcomes of previous plays of the same game by other players; it might give a background story that would tend to identify some particular division of the terrain as corresponding to an original "status quo"; or it might seem to attach a kind of moral claim of one of the players to particular parts of the board. These background data would have no influence on the logical or mathematical structure of the game; they would be intended to have no force except power of suggestion. Again, one might set up the board so that on the first play

[3] A splendid example of the creation of norms in practice — and one that suggests that the process is susceptible of analysis — was the rather general acceptance during the 1957 disarmament discussions of the notion that any inspection zone ultimately agreed on had to be selected from among the array of possible pie-shaped zones with apex at the North Pole.

[4] One may hope, as a game theorist, that a clear line can be drawn between the experimental psychology pertinent to game theory and the rest of social psychology; this is still supposed to be a theory of *strategy*, not the entire domain of conflict behavior. But it is not clear just where the line can be drawn in advance. "Hostility," for example, might seem to be an emotional or temperamental quality best kept out of game theory; but if a player's hostility in the game is a significant constraint on his ability to perceive the other player's meaning, it becomes part of the "communication structure." An experiment by Deutsch is pertinent. He let pairs of players play nonzero-sum games (in matrix form) tacitly for a sequence of two plays, the game providing both a "cooperative" and an "uncooperative" choice. Those who played uncooperatively against a cooperative partner had an opportunity, on the second play, to respond to the implicit offer of cooperation. But, "when their expectation of the other person's choice was not confirmed, they tended to interpret his choice as being a function of indifference or a basic lack of understanding as to how the game 'should' be played. . . . In this group, knowledge of the other person's choice, because of the meaning attributed to it, tended to reinforce the previous negative sentiments regarding the intentions of the other person." See Morton Deutsch, *Conditions Affecting Cooperation*, Research Center for Human Relations, New York University, 1957. (An article based on this monograph, not including the point quoted here, entitled "Trust and Suspicion," appeared in *The Journal of Conflict Resolution*, 2:265–279 [December 1958].)

it corresponds to the way it stood in the middle of the same game as played earlier by two other players, and see whether the outcome can be affected by informing the players of what the starting lineup was in that earlier game. If players tend to develop "norms" based on the static configuration of the game as they appreciate it at the outset, it may be possible to distort those norms by providing, in a completely "nonauthoritative" way, a background story that suggestively indicates some other hypothetical starting point.[5]

It should also be interesting to see whether each player can really discern when the other is "testing" his determination, "daring" him, and so forth; and it might be possible to study the process by which particular encounters become invested with symbolic importance, such that each player recognizes that he is establishing a role and reputation in the way he conducts himself at a particular point in the game.

Another dimension of the game that seems susceptible of analysis is the significance of the *incrementalism* that is involved in the moves and value systems. Take, for example, a game that involves moving pieces over a board or troops over some terrain. If players move in turn, each moving one piece one square at a time, the game proceeds at a slow tempo by small increments; the situation on the board may change character in the course of play, but it does so by a succession of small changes that can be observed, appreciated, and adapted to, with plenty of time for the mistakes of individual players or mutual mistakes that destroy value for both of them to be observed, adapted to, and avoided in subsequent play. If there is communication, there is time for the players to bargain verbally and to avoid moves that involve mutual destruction. But suppose that, instead, the pieces can be moved several at a time in any direction and any distance and that the rules make the outcome of any hostile clash enormously destructive for one or both sides. Now the game is not so incremental; things can happen abruptly. There may be a temptation toward surprise attack. While one can see what the situation is at a particular moment, he cannot project it more

<hr />

[5] The income-tax questions described in Chapter 3 (pp. 62–65) indicate the force of this power of suggestion.

than a move or two ahead. There seems to be less chance to develop a modus vivendi, or tradition of trust, or dominant and submissive roles for the two players, because the pace of the game brings things to a head before much experience has been gained or much of an understanding reached. But does a more incremental game make successful collaboration easier, or does it just invite a riskier mode of play? Or does this depend on what kinds of people the players are and on what suggestions we plant in the game itself? Is the critical factor the incrementalism of the *moves* in the game or incrementalism in the *value* systems of the players (that is, of the scoring system)? Or can these be made commensurate with each other, so that incrementalism can be introduced into a game in one dimension to offset the lack of it in another? The relevance of these questions is attested by the controversy over the role of nuclear weapons in limited war, the significance of the temptation to surprise attack in a situation that depends on mutual deterrence, and various proposals to reduce the tempo of modern war and to isolate it geographically, together with disagreement over whether there can be such a thing as limited war on the continent of western Europe. Incrementalism may be comparatively amenable to formal analysis, once the necessary empirical benchmarks have been identified by experiment or observation.[6]

These questions have concerned two-person games, except for the possible role of the mediator. Similar games could be played by three or more participants, each on his own account; and the author conjectures that — at least among "successful" players — many of the empirical results would appear in sharper relief with the larger number of players. More generally, the kind of coordination involved in the formation of mobs and coalitions may lend itself to experimental study. In contrast to the more sanitary, symmetrical schemes that have sometimes been used to

---

[6] "It is not only that limited war must find means to prevent the most extreme violence; it must also seek to slow down the tempo of modern war lest the rapidity with which operations succeed each other prevent the establishment of a relation between political and military objectives. If this relationship is lost, any war is likely to grow by imperceptible stages into one all-out effort" (Henry A. Kissinger, *Nuclear Weapons and Foreign Policy* [New York, 1957]).

study the formation of coalitions in game theory, it might prove more interesting to introduce deliberately certain asymmetries, precedents, orders of moves, imperfect communication structures, and various connotative details, in order to study the crystallization of groups. Certainly the influence exerted on the formation of coalitions by various kinds of asymmetrical and otherwise imperfect communication systems often lends itself to systematic experimental study.[7]

[7] Alex Bavelas has described an experiment in pure coordination in which each of five separated players must pass geometric pieces among themselves until they reach a distribution of the pieces that permits the formation of five separate squares. The pieces are so cut that many "wrong" squares can be formed, that is, squares that use a combination of pieces that makes it impossible for four more squares to be formed with the remaining pieces. He is interested in what happens when these deceptive "successes" occur. "For an individual who has completed a square it is understandably difficult to tear it apart. The ease with which he can take a course of action 'away from the goal' should depend to some extent upon his perception of the total situation. In this regard the pattern of communication should have well-defined effects. . . . Preliminary runs . . . have revealed . . . that the binding forces against restructuring are very great, and that, with any considerable amount of communication restriction, a solution is improbable" ("Communication Patterns in Task-oriented Groups," in D. Cartwright and A. F. Zander, *Group Dynamics* [Evanston, 1953], p. 493). Some very suggestive experimental work, especially on "the biased perception of what is equable," is reported by Charles E. Osgood, "Suggestions for Winning the Real War with Communism," *Journal of Conflict Resolution*, 3:304–05 (December, 1959).

PART III

STRATEGY WITH A
RANDOM INGREDIENT

7

RANDOMIZATION OF PROMISES
AND THREATS

In the theory of games of pure conflict (zero-sum games) randomized strategies play a central role. It may be no exaggeration to say that the potentialities of randomized behavior account for most of the interest in game theory during the past one and one-half decades.[1] The essence of randomization in a two-person zero-sum game is to preclude the adversary's gaining intelligence about one's own mode of play — to prevent his deductive anticipation of how one may make up one's own mind, and to protect oneself from tell-tale regularities of behavior that an adversary might discern or from inadvertent bias in one's choice that an adversary might anticipate. In the games that mix conflict with common interest, however, randomization plays no such central role, and the role it does play is rather different.[2]

[1] John von Neumann, speaking of "the fundamental theorem on the existence of good strategies," namely the theorem that all zero-sum games with a finite number of pure strategies have a minimax-maximin equilibrium pair ("solution") if mixed strategies are allowed, said, "As far as I can see, there could be no theory of games on these bases without that theorem. . . . Throughout the period in question I thought there was nothing worth publishing until the 'minimax theorem' was proved" ("Communication on the Borel Notes," *Econometrica*, 21:124–125 [January 1953]).

[2] One can, instead, interpret mixed strategies in zero-sum games as a means of introducing continuity of strategies into a discrete-strategy game that has no pure-strategy saddle point, thereby converting it into a game that does have a saddle point. In this interpretation the role of mixed strategies in zero-sum games is not so different from their role in the nonzero-sum games. One can flip a coin to keep an opponent from guessing with confidence whether it will come up heads or tails; or one may flip a coin to "average" heads and tails, to create (in an expected-value sense) a strategy halfway between heads and tails. Both interpretations are useful. If the second is somewhat more sophisticated, the first may better catch the spirit of the problem as it presents itself to a game player. And the first reminds us that the problem, even with

Randomization in the theory of these ("nonzero-sum") games is not mainly concerned with preventing one's strategy from being anticipated. In these games, as noted earlier, one is often more concerned with making the other player anticipate one's mode of play, and anticipate it correctly, than with disguising one's strategy.

There may of course be zero-sum components embedded in a larger game. In limited war one may be concerned to communicate rather than to disguise the limits that one proposes to observe, but within those limits may sortie his aircraft in a randomized way to minimize the enemy's tactical intelligence.[3] Again, information samples may be exchanged, or agreements enforced on a sample basis, where neither party can afford to yield the other full knowledge. Arms-control agreements, for example, might have to be monitored by a sampling technique that yielded each side enough knowledge about the enemy's forces to reveal compliance or noncompliance without yielding so much that the possibility of successful surprise attack on those forces were greatly enhanced.

But the main role of randomization in the traditional literature on nonzero-sum games is a different one. It has been a device to make indivisible objects divisible, or incommensurate objects homogeneous. Their "expected values" are divisible by lottery when the objects themselves are not. We flip coins to see who

---

randomization, is still to prevent the opponent's anticipation of our actual strategy choice, and that the machinery of choice, the procedures for recording and communicating a choice, and any advance preparations required by the outcome of the random process, must remain inaccessible to his intelligence system.

[3] In particular cases there may be a tantalizing dilemma inherent in a choice of secrecy or revelation. If in order to prove that one is committed to a threat, or that one is in fact capable of fulfilling the threat, one must display evidence of the commitment or the capability to the other party, the evidence may be of a kind that necessarily yields information helpful to the second party in combatting the threat. To prove to an enemy that one has a potent weapon that can overcome his defenses we might have to demonstrate the weapon or some aspect of it, or provide technical knowledge to prove the weapon feasible; to do so may aid him greatly in preparing a defense against it. If, to prove we would fight a local war in an ambiguous area, it were necessary to station troops there ahead of time, the enemy would have the advantage of knowing their exact location rather than having to be prepared in all directions.

gets the object, and play "double or nothing" when we cannot make change. We can divide the obligation of citizenship equally by selecting draftees through a lottery, when we want a fraction of the eligibles for a long period of service rather than all of them for a short one.

In this role, randomization is evidently relevant to promises. If the only favors available to be promised are larger than necessary and not divisible, a lottery that offers a specified probability of the favor's being granted can scale down the expected value of the promise and reduce the cost to the person making it. An offer to help a person on a large scale in a contingency is somewhat equivalent to offering the certainty of smaller help. (There may be the additional advantage that the contingency is correlated with his need.)

But in this respect a promise is different from a threat. The difference is that a promise is costly when it succeeds, and a threat is costly when it fails. A successful threat is one that is not carried out. If I promise more than I need to as an inducement, and the promise succeeds, I pay more than I needed to. But a threat that is "too big" is likely to be superfluous rather than costly. If I threaten to blow us both to bits when it would have been sufficient to threaten our discomfort, you'll likely still comply; since I have neither to discomfort us nor to kill us, the error costs nothing. If all I had was a grenade to explode in our midst and wished for tear gas instead, I might scale down the grenade to the "size" of a tear-gas bomb by threatening an appropriate percentage chance that the bomb would go off, killing us both, if you failed to comply. But the need to do this is not as clear as in the case of a promise, where any excess in the value promised is so much loss.

The size of the threat can be a problem if it costs something to be equipped to *make* a threat and if bigger threats cost more to make than small ones. If a threat of tear gas is enough, so that I do not need to threaten explosion, and if tear-gas bombs are cheaper than explosive ones, and if I have to display the bomb to make the threat persuasive, it is better to threaten with the cheaper tear gas. But grenades may be cheaper, and then the incentive goes the other way. For many interesting threats the greatest cost is the risk of having to carry it out, and the more ordinary "cost" is not a controlling factor.

THE RISK OF FAILURE

The risk of *failure*, however, does give an incentive to choose moderate rather than excessive threats. If the only threat that can be made is some horrendous act, one may be tempted to scale it down by attaching it to a lottery device — by threatening some *specified probability* that it will be carried out unless compliance is forthcoming, not by committing oneself to the certainty that the jointly painful punishment would be administered.

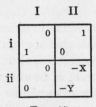

FIG. 17

To illustrate, consider the matrix in Fig. 17, in which Column has first choice, followed by Row, but in which Row has the option of making a prior threat to constrain Column's choice. (Interpret X and Y as positive numbers.) On one condition, Row's strategy is clearly to threaten row ii if Column chooses column II. If he makes no threat, Column chooses II knowing that Row will then choose i. Given the threat — and assuming that Row is committed to it and that Column knows it — the choice of II yields unattractive outcomes for both of them, and Column can be expected to choose I.

The condition is that Row be quite sure that nothing will go wrong! Maybe he completely misjudges Column's payoffs. Maybe this particular adversary is drawn from a universe in which nearly everyone, but not quite everyone, has preferences as indicated in the matrix, and a few deviants have a radically different preference system and prefer the lower right cell to the upper left one. Alternatively, Row may get himself committed to his threat but fail to communicate it convincingly to Column, so that Column mistakenly ignores the threat, condemning them both to the lower right-hand cell. Again, Column himself may have arranged a prior commitment through his own choice of II, and failed to communicate it accurately to Row in time for Row to take this into account, or Column may have suffered a disability unknown to Row that eliminates the possibility of I; in that case, Row's own commitment will only guarantee the worst outcome for both players. Whatever the reasons for failure, there is perhaps some probability that the threat will fail. If we take it into account we may have a reason for Row to wish that the "punitive" payoffs in the lower right-hand cell were not quite as unattractive as they are.

If Row is confined to "pure" strategies — if he must specify his threat or commitment without reference to error or chance — he can do nothing but wish that the numbers in the lower right-hand cell were not so unattractive. But if he can randomize his threat he can in fact "scale it down" to reduce somewhat the high cost of failure. If, for example, he can commit himself not to a choice of row ii in the event that column II is chosen, but to a 50-50 chance between i and ii in that event, he may still hope to frighten Column into a choice of I while reducing the seriousness of the risk of failure.

We can be more specific. Let P stand for the probability that the threat will fail for any reason whatsoever. (For our present purpose this is an "autonomous" probability, independent of Row's strategy.) Let Row now threaten to choose ii with probability equal to π, in the event Column chooses II. In other words, if Column fails to comply there is a probability of π that Row will choose ii to their mutual discomfort, and of $(1 - \pi)$ that he will choose i to their mutual relief. What value of π should Row choose?

First, how large does π have to be to make the threat effective at all, that is, to make it effective assuming that it does *not* fail for any of the autonomous reasons involved in P? This is a question of Column's choice when he is confronted with the risk π. If Column chooses I he gets 0. If he chooses II his expectation is a weighted average of 1 and $-X$, with weights of $(1 - \pi)$ and π

respectively. If this average is less than o, he is motivated to choose I — subject to the autonomous probability, P, that for one reason or another he will choose II in spite of his apparent motivation toward I. The condition for an effective threat is thus [4]

$$o > (\mathrm{1} - \pi) - \pi X,$$

$$\pi > \frac{\mathrm{1}}{\mathrm{1} + X}.$$

Second, assume that any threat with π above the floor established by the preceding formula will succeed or fail with probabilities $(\mathrm{1} - P)$ and P respectively. If the threat succeeds, Row's payoff is $+\mathrm{1}$. If it fails, his expectation is a weighted average of o and $-X$, the weights being $(\mathrm{1} - \pi)$ and π respectively. The expected value of the outcome, then, when the threat is large enough to be effective at all, is given by

$$(\mathrm{1} - P) + P(o - \pi Y) = \mathrm{1} - P - P\pi Y.$$

This value is evidently higher, the lower is the value of π. Row should therefore arrange the lowest value of π that he can that meets the first condition. For a threat to be worthwhile at all — to have an expected value greater than zero, which is what Row can expect from this particular matrix if he makes no threat — a value of π must be arranged that meets the condition

$$\mathrm{1} - P - P\pi Y > o$$

or

$$\frac{\mathrm{1} - P}{P} \cdot \frac{\mathrm{1}}{Y} > \pi.$$

Thus the effective range for π in this example is given by

$$\frac{\mathrm{1} - P}{P} \cdot \frac{\mathrm{1}}{Y} > \pi > \frac{\mathrm{1}}{\mathrm{1} + X}.$$

And there is no threat at all worth making if there is no room between these two limits, if

---

[4] Since the analysis depends only on comparisons of the *differences* between absolute valuations of the payoffs for the two players separately, no violence is done by adopting, for each player, a scale of measurement that sets his preferred payoff equal to $+\mathrm{1}$ and his next preferred payoff to o. The full interpretation, then, of the expression $\mathrm{1}/(\mathrm{1} + X)$, is: the ratio of (1) the difference between Column's upper right and upper left payoffs, to (2) the sum

or
$$\frac{\mathrm{I} - P}{PY} < \frac{\mathrm{I}}{\mathrm{I} + X}$$

$$\frac{P}{\mathrm{I} - P} > \frac{X + \mathrm{I}}{Y}.$$

Only a "fractional" threat — a threat with π less than I — is worth making if:

or
$$\frac{\mathrm{I} - P}{PY} < \cdot$$

$$\frac{P}{\mathrm{I} - P} > \frac{\mathrm{I}}{Y}.$$

Here is a case, then, in which the fractional threat is superior to the certainty threat, and in which the latter could be not worth making at all while the former were. The argument hinges on the risk of failure, a risk that has been assumed independent of the size of π itself. This is a somewhat special assumption. If we interpret P as the probability that we have misjudged our adversary and exaggerate his preference for avoiding the lower right cell, our assumption implies a bimodal distribution of payoffs in the population. It implies that we have either a man whose payoffs are adequately represented by the numbers in our matrix, or a man whose payoffs are so different that no relevant threat — within the range of values up to $\pi = \mathrm{I}$ — will dissuade him. If instead we supposed that the ratio of column payoffs in the upper and lower right-hand cells showed a bell-shaped frequency distribution within the population, and that our particular adversary had been drawn at random, the probability that our threat would succeed would vary directly with

of the differences between (a) his upper right and upper left payoff and (b) his lower right and upper left payoffs. The simplicity of the formulae thus reflects advantage already taken of this scaling convenience. It takes only one parameter to characterize the relevant relations among three valuations. (In a later problem that involves the lower left cell, all four payoffs are relevant and a second parameter would be required. That case, however, can be further simplified if the lower left payoff can be taken equal to one of the others and still illustrate the point; we get less complete knowledge but more o's and I's that way.) On the interpretation of these numbers see A. A. Alchian, "The Meaning of Utility Measurement," *American Economic Review*, 43: 26–50 (March, 1953), or Luce and Raiffa, pp. 12–38.

the value of π itself. The probability that a burglar drawn at random from the universe of burglars will be deterred by some specified probability of apprehension and conviction presumably varies directly with the latter probability; the simple model analyzed above treats burglars as divisible into two classes — those, let us say, who steal for money and are certainly deterred in accordance with the numbers of the matrix, and those who steal for fun and are beyond reach of any threat of the magnitude entered in the lower right-hand cell. On the other hand, if our probability of failure reflected, say, a breakdown of communication with the adversary, there might be better reason for supposing the probability of failure to be independent of the particular threat being communicated.

It is interesting to notice that attaching a probability of fulfillment to our threat is, in the above model, substantially equivalent to scaling down the size of the threat more directly. To see this, interpret X in the lower right-hand cell as a fine that will be levied on both Row and Column, or a number of lashes with the whip or days of imprisonment that both will suffer if the threat is fulfilled. If X is the maximum number of dollars, lashes or days that Row can threaten, let π be interpreted as Row's specification of what fraction of the maximum permissible penalty is to be exacted; if π is set at 0.5, for example, both Row and Column receive exactly half their maximum punishments. If we interpret the matrix in this way, and ask what value of π provides the optimum threat from Row's point of view, we go through the same analysis and we reach the same conclusion as before, namely, π is to be as small as possible subject to a minimum value equal to $1/(1 + X)$. Thus we can interpret π either as a probability of threat fulfillment or as the scale on which the threat is to be certainly carried out. Since the two formulations come to the same thing, and we can interpret π either way, it seems fair to say that *in this case* the role of randomization is that of making divisible an otherwise too large and indivisible threat, of making possible a "smaller" threat than was otherwise available. (It should be noted though, that to reduce a threat by reducing the probability of its fulfillment reduces the expected value of the outcome proportionately for both players, while a direct reduction in size might not be restricted to proportionate changes in value or utility for the two parties.) [5]

THE RISK OF INADVERTENT FULFILLMENT

There is another "cost" element that can motivate a reduced threat. This is the risk that one will fulfill the threat inadvertently, even if the adversary does comply with it (or would have complied if the threat hadn't gone off accidentally before he had a chance). The gun that threatens a burglar or hold-up victim may go off accidentally before he has a chance to comply. The dog that threatens to bite trespassers may bite some who do not trespass.

If a hitchhiker pulls a gun on the driver of a car and the driver threatens to kill them both unless the hitchhiker throws his gun out the window, making his threat by pressing the accelerator to the floor and creating a manifest risk of fatal accident, there is some chance that the accident will occur before the hitchhiker has a chance to comprehend the threat and comply. In this case, the risk of accidental fulfillment is an integral part of the threat. The only way one can make the threat is to start fulfilling it. Until the driver speeds up the hitchhiker has no reason to believe him; once he does speed up, there is some minimum length of time it takes the hitchhiker to comply and the driver to relax his speed. There is therefore an interval, however short it may be, that the risk is present; the risk entailed by the high speed must therefore be one that is small enough to be tolerable to the driver during this initial interval. If instead the car were definitely safe at all speeds under sixty but would certainly skid off the road at exactly sixty and there were no gradations between that carried a moderate risk of accident, the driver could have no incentive to incur a dangerous speed and the hitchhiker would know it and not respond to a verbal threat of high speed. It is the possibility of a "fractional threat," a threat that carries the risk but not the certainty of death, that gives the driver

[5] Randomization may also be integrally related to the arrangement of the threat itself, or be involved in the decision process whether the threatener wishes it or not. So the interpretation of randomization as just a means of manipulating the size of the threat is applicable only in some cases.

anything to work with; but to put it into effect he has to suffer it for some finite period.

If in situations of this kind we suppose — as is roughly true in the hitchhiker case — that the risk of inadvertent fulfillment is proportionate to the probability, π, that one will fulfill the threat if the adversary does not comply — if the watchdog's propensity to bite innocent passersby is proportionate to his proclivity to bite those who enter the premises — a formula is obtained that is not very dissimilar to the one already arrived at. Using the same matrix as before (ignoring this time the probability that a potentially effective threat may fail) and letting $a\pi$ represent the probability of inadvertent fulfillment, the minimum value of π is the same as before. The expected value of the outcome to Row, which must exceed o if he is to make the threat, is given by the left-hand side of the formula

$$(1 - a\pi) - a\pi Y > 0,$$

or

$$\frac{1}{a(1 + Y)} > \pi > \frac{1}{1 + X}.$$

The optimal threat is again one that barely exceeds the lower limit; there is an upper limit to π that may be less than 1; and, depending on the relative values of X and Y and the "cost" parameter a, it may or may not be possible to find a profitable value for π at all.

RANDOMIZED COMMITMENTS

Having found a rationale for a "fractional threat," we can inquire whether the tactic of "unconditional commitment," too, is one that in certain cases can advantageously be made less than certain. As indicated in Chapters 3 and 5,[6] a pure commitment — that is, a definite commitment to a pure strategy — is equivalent to "first move" in a two-person, two-move game in which one would otherwise have to move second; it is a means of obtaining the equivalent of first move. We have to relax that interpretation if we suppose that Row, who has second move in the game but who has the option to commit himself ahead of time, commits himself to a 50-50 chance of choosing row i or ii. To

[6] Pp. 47, 122.

do this one must retain the right to move second, exploiting only the right to commit oneself ahead of time; if one had actually to move first, by a definite choice, the possibility of a randomized commitment would be lost. (The randomized commitment is equivalent to a "first move" determined by a random device with odds set by the player, with the odds but not the actual move known to the other player before his own move.)

The same payoff matrix (Fig. 1) can be used to illustrate this situation if we change the rules of the game to permit Row an *unconditional* commitment prior to Column's choice but not permitting him to make his choice depend on Column's. A firm commitment to ii induces a choice of column I but is wasted because the lower left cell — to which Row is now committed — contains no reward. Row's problem is that he needs row ii to induce Column into I, but he needs row i to profit from I. A compromise can be achieved by a randomized commitment — a commitment to a randomized choice. If Row is committed to flip a coin (50-50 chance) to select i or ii after Column has chosen, Column will choose I as long as X is greater than 1.[7] In that case Row gets an expected value of 0.5. If Row sets π (the probability of his choosing ii) at just above $1/(1 + X)$ he gets the largest expected value consistent with Column's choice of I. (If Column's payoff in the lower left cell differs from zero, say 0.5 or −0.5, the formula for optimum value of π differs somewhat.) If Row's payoff in the lower left cell were −1, no commitment with a greater than 50 per cent chance of ii would serve. And if that payoff were −X or worse, no probability mixture of i and ii would work; any mixture with π large enough to induce column I would be too large to yield Row a positive expected value.

There is another rationale for a fractional commitment. In the case just discussed, it was Row's own preference for the upper cell in I that led him to minimize the value of π. In Fig. 18 it is *Column's* motivation that demands some chance of row i, that is, a fractional value of π. In this case, a firm commitment to row ii induces Column to choose II; a firm commitment to i induces

---

[7] That is, as long as the payoff to Column in the lower right cell falls short of his payoff in the upper left as much as the payoff in the upper right exceeds the upper left. See the earlier footnote on the scaling of payoffs.

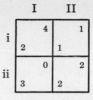

FIG. 18

Column to choose I; no commitment at all leaves Column preferring II; a threat to choose i unless Column chooses I will be ineffective unless Row promises to abstain from choosing ii. In all of these "pure-strategy" cases, Row ends up with a score of 2. He can, however, do slightly better with a mixed commitment. He can, because he and Column are both attracted to column I, disagreeing only over the choice of Row in that column. If he offers Column a 50-50 chance between rows i and ii, Column gets an expected value of 2 in the first column, of 1.5 in the second, and chooses the first. This leaves Row an expected value of 2.5. Since Row has a preference for ii, he wants the highest probability of that row consistent with the need to provide Column with a preference for column I. That is, he wants the largest value of π for which (in the matrix shown)

$$4(1 - \pi) > (1 - \pi) + 2\pi$$
or
$$3/5 > \pi.$$

This particular mixed commitment can be called a *combination* of a fractional threat with a fractional promise. Row, in effect, "threatens" a relatively high probability of i in the event that II is chosen and "promises" it if I is chosen.

He could do even better if he could make π *conditional* on Column's choice. Any probability up to 0.75 for row ii, conditional on a choice of column I, is a sufficient inducement if it is certain that Row will retaliate for column II with row i. But if he is limited to making his threat no worse than his promise is good — if he has to attach the same probability to both of them — the upper limit to an effective value of π is 0.6, with an expected value to Row of 2.6 (and of 1.6 for Column). With a separate π for the promise, the upper limit is 0.75 for an expected payoff of 2.75 (and only 1.0 for Column).

8

THE THREAT THAT LEAVES
SOMETHING TO CHANCE

It is typical of strategic threats that the punitive action — if
the threat fails and has to be carried out — is painful or costly
to both sides. The purpose is deterrence *ex ante*, not revenge *ex
post*. Making a credible threat involves proving that one would
have to carry out the threat, or creating incentives for oneself or
incurring penalties that would make one evidently want to. The
acknowledged purpose of stationing American troops in Europe
as a "trip wire" was to convince the Russians that war in Europe
would involve the United States whether the Russians thought
the United States wanted to be involved or not — that escape
from the commitment was physically impossible.

As a rule, one must threaten that he *will* act, not that he *may*
act, if the threat fails. To say that one *may* act is to say that one
may not, and to say this is to confess that one has kept the
power of decision — that one is not committed. To say only that
one *may* carry out the threat, not that one certainly will, is to
invite the opponent to guess whether one will prefer to punish
himself and his opponent or to pass up the occasion. Furthermore, if one says that he may — not that he will — and the opponent fails to heed the threat, and the threatener chooses not to
carry it out, he only confirms his opponent's belief that when he
has a clear choice to act or to abstain he will choose to abstain
(consoling himself that he was not caught bluffing because he
never said that he would act for sure).

There are threats of this kind nevertheless that may be effective in spite of this loophole. They can work, however, only
through a process that is a degree more complicated than firm

commitment to certain fulfillment. Furthermore, they may arise inadvertently and may entail unintended behavior. For this reason they are less likely to be recognized and understood.

The key to these threats is that, though one may or may not carry them out if the threatened party fails to comply, *the final decision is not altogether under the threatener's control.* The threat is not quite of the form "I may or may not, according as I choose," but, has an element of, "I may or may not, and even I can't be altogether sure."

Where does the uncertain element in the decision come from? It must come from somewhere outside of the threatener's control. Whether we call it "chance," accident, third-party influence, imperfection in the machinery of decision, or just processes that we do not entirely understand, it is an ingredient in the situation that neither we nor the party we threaten can entirely control. An example is the threat of inadvertent war.

THE THREAT OF INADVERTENT WAR

The thought that general war might be initiated inadvertently — through some kind of accident, false alarm, or mechanical failure; through somebody's panic, madness, or mischief; through a misapprehension of enemy intentions or a correct apprehension of the enemy's misapprehension of ours — is not an attractive one. As a general rule one wants to keep such a likelihood to a minimum; and on the particular occasions when tension rises and strategic forces are put on extraordinary alert, when the incentive to react quickly is enhanced by the thought that the other side may strike first, it seems particularly important to safeguard against impetuous decision, errors of judgment, and suspicious or ambiguous modes of behavior. It seems likely that, for both human and mechanical reasons, the probability of inadvertent war rises with a crisis.

But is not this mechanism itself a kind of deterrent threat? Suppose the Russians observe that whenever they undertake aggressive action tension rises and this country gets into a sensitive condition of readiness for quick action. Suppose they believe what they have so frequently claimed — that an enhanced status for our retaliatory forces and for theirs may increase the danger of an accident or a false alarm, theirs or ours, or of some triggering incident, resulting in war. May they not perceive that the risk of all-out war, then, depends on their own behavior, rising when they aggress and intimidate, falling when they relax their pressure against other countries?

Notice that what rises — as far as *this* particular mechanism is concerned — is not the risk that the United States will *decide* on all-out war, but the risk that war will occur whether intended or not. Even if the Russians did not expect deliberate retaliation for the particular misbehavior they had in mind, they could still be uneasy about the possibility that their action might precipitate general war or initiate some dynamic process that could end only in massive war or massive Soviet withdrawal. They might not be confident that we and they could altogether foretell the consequences of our actions in an emergency, and keep the situation altogether under control.

Here is a threat — if a mechanism like this exists — that we *may* act massively, not that we certainly will. It could be most credible. Its credibility stems from the fact that the possibility of precipitating major war in response to Soviet aggression is not limited to the possibility of our coolly deciding to attack; it therefore extends beyond the areas and the events for which a more deliberate threat is in force. It does not depend on our preferring to launch all-out war, or on our being committed to, in the event the Russians confront us with the *fait accompli* of a moderately aggressive move. The final decision is left to "chance." It is up to the Russians to estimate how successfully they and we can avoid precipitating war under the circumstances.

The threat — if we call this contingent-behavior mechanism a "threat" — has some interesting features. It may exist whether we realize it or not. Even those who have doubted whether our massive-retaliation threat was a potent deterrent to *minor* aggression during the last several years, but are perplexed that the Russians have not engaged in more mischief than they have, can note that the threat we voiced was backed by an additional implicit threat that we might be triggered by Soviet actions in spite of ourselves. Furthermore, even if we prefer not to incur even a small probability of inadvertent war, and would not use this mechanism deliberately, the "threat" in question may be a by-product of other actions that we have a powerful incentive to take. We may get this threat whether we like it or not when we (and the Russians) take precautions commensurate with a crisis; knowing this, the Russians may have to take the risk into account. Finally, the threat is not discredited even if the Russians accomplish their purpose without triggering war. If the Russians estimate that the chance of inadvertent war during a particular month rises from very small to not-so-small if they create a crisis, and they go ahead anyway, and no major war occurs, they still have little reason to suppose that their original estimate was wrong, and little reason to suppose that repetition would be less risky, any more than a person who survives a single play of Russian roulette should decide it isn't dangerous after all.

LIMITED WAR AS A GENERATOR OF RISK

Limited war as a deterrent to aggression also requires interpretation as an action that enhances the *probability* of a greater war. If we ask how the Western forces in Europe are expected to deter a Russian attack or to resist it if it comes, the answer usually runs in terms of a sequence of *decisions*. In case of attack on a moderate scale, we could make the decision to fight limited war; it would not be a decision to proceed with mutual annihilation. If we can resist the Russians on a small scale, they must either give up the idea or themselves take a step upward on the scale of violence. At some point there is a discontinuous jump from limited war to general war, and we hope to confront *them* with that choice. If this is not the typical sequence of decisions envisaged, it at least seems typical in one respect: it involves *deliberate* decisions — decisions to take an action or to abstain from it, to initiate a war or not to, to step up the level of violence or not to, to respond to a challenge or not to.

But another interpretation can be put on limited war. The danger of all-out war is almost certainly increased by the occurrence of a limited war; it is almost certainly increased by an enlargement of limited war. This being so, the threat to engage in

limited war has two parts. One is the threat to inflict costs directly on the other side, in casualties, expenditures, loss of territory, loss of face, or anything else. The second is the threat to expose the other party, together with one's self, to a heightened risk of general war.[1]

Here again is a threat that all-out war *may* occur, not that it certainly will occur, if the other party engages in certain actions. Again, whether it does or does not occur is not a matter altogether controlled by the threatener. Just how all-out war would occur — just where the fault, initiative, or misunderstanding may occur — is not sure. Whatever it is that makes limited war between great powers a risky thing, the risk is a genuine one that neither side can altogether dispel if it wants to. The final decision, or the critical action that initiates an irreversible process, is not something that should necessarily be expected to be taken altogether deliberately. "Chance" helps to decide whether general war occurs or not, with odds that are a matter of judgment based on the nature of the limited war and the context in which it occurs.

Why would one threaten limited war rather than all-out war to deter an attack? First, to threaten limited war — according to this analysis — is to threaten a risk of general war, not the certainty of it; it is consequently a lesser threat than the massively retaliatory threat and more appropriate to certain contingencies. Second, it has the advantage, in case the enemy misjudges our intentions or commitments, of an intermediate stage: we can *engage* in limited war, creating precisely the risk for both of us that we threatened to create, without thereby making general war the price we both pay for the enemy's mistaken judgment. We pay instead the lesser price of a risk of general war, a risk that the enemy can reduce by withdrawal or settlement.

Third, in case the enemy is irrational or impetuous, or we have misjudged his motives or his commitments, or in case his aggressive action has gotten up too much momentum to stop, or his actions are being carried out by puppets or satellites that are

[1] The same point is stressed by Glenn H. Snyder, "Deterrence by Denial and Punishment" (Research Monograph No. 1: Princeton University Center of International Studies, January 2, 1959), pp. 12, 29.

beyond his immediate power to control, there is some prudence in threatening risk rather than certainty. If we threaten all-out war, thinking it not too late to stop him, and it is, we must either go ahead with it or have our threat discredited. But if we can threaten him with a one-in-twenty chance of all-out war in the event he proceeds, and he does proceed, we can hold our breath and have nineteen-to-one odds of getting off without general war. Of course, if we scale down the risk to us, we scale it down to him too; it may degrade the threat to put too much safety in it. But in cases where there is danger that we completely misjudge the enemy's commitment to an action, or completely misjudge his ability to control his own agents, allies, or commanders, the more moderate risk may deter anything that is still within his control.

If we give this interpretation to limited war, we can give a corresponding interpretation to enlargements, or threats of enlargement, of the war. The threat to introduce new weapons into a limited war is not, according to this argument, to be judged solely according to the immediate military or political advantage, but also according to the deliberate risk of still larger war that it poses. Just as a moderate limited war may increase by a large factor the likelihood of major war within the next thirty days, so a progression from conventional to novel weapons may raise that probability by another factor.

We are led in this way to a new interpretation of the "trip wire." The analogy for our limited-war forces in Europe is not, according to this argument, a trip wire that certainly detonates all-out war if it is in working order and fails altogether if it is not. What we have is a graduated series of trip wires, each attached to a chance mechanism, with the daily *probability* of detonation increasing as the enemy moves from wire to wire. The critical feature of the analogy, it should be emphasized, is that whether or not the trip wire detonates general war is — at least to some extent — outside our control, and the Russians know it.

The same interpretation might be true of Quemoy. One can argue that the Chinese or Russians were deterred by the prospect of major war, not just by the prospect of losing a limited war or winning one at excessive cost. Even if they were convinced that we would exercise every skill and caution to keep a war limited, and they were prepared to exercise skill and caution themselves, they may simply have felt that the process that leads to bigger and bigger wars is not one that they or we fully understand or can foresee, and that the risk, though numerically small, was appreciable.

RISKY BEHAVIOR IN LIMITED WAR

If one of the functions of limited war, then, is to pose the deliberate risk of all-out war, in order to intimidate the enemy and to make pursuit of his limited objectives intolerably risky to him, the usual precepts for behavior in limited war need revision. The supreme objective may not be to *assure* that it stays limited, but rather to keep the risk of all-out war within moderate limits *above zero*. At least this may be the strategy for the side that is in danger of "losing" a limited war. The less likely it is that the enemy's aggressive advances can be contained by limited and local resistance, the more reason there may be to fall back upon the deliberate creation of mutual risk. (Alternatively, the more the aggressor can design his advances so that even local resistance seems fraught with explosive potential, the less attractive local resistance will seem.)

Deliberately raising the risk of all-out war is thus a tactic that fits the context of limited war. Of course, one cannot raise the risk just by saying so. One cannot just announce to the enemy that yesterday one was only about 2 per cent ready to go to all-out war but today it is 7 per cent and they had better watch out. One has to take actions that — assuming he and his adversary continue to be just as concerned and careful to keep the war limited — leave everyone just a little less sure that the war can be kept under control.

The idea is simply that a limited war can get out of hand by degrees. At any point one has some notion or sensation of how much "out of control" it is. And various actions — innovations, breaches of limits, manifestations of "irresponsibility," challenging and assertive acts, adoption of a menacing strategic posture, adoption of headstrong allies and collaborators, spoofing and harassing tactics, introduction of new weapons, enlargement of troop commitments or the area of conflict — tend to raise almost anyone's judgment of how much "out of control" the situation is. To share such an increase in risk with an enemy may provide him an overpowering incentive to lay off. Preferably one creates the shared risk by irreversible maneuvers or commitments, so that only the enemy's withdrawal can tranquilize the situation; otherwise it may turn out to be a contest of nerves.

REPRISAL AND HARASSMENT

Limited local war is not the only context in which deliberately risky behavior may be used as a type of threat. Between the threats of massive retaliation and of limited war there is the possibility of less-than-massive retaliation, of graduated reprisal. Few serious analyses of war of limited reprisal have been published.[2] The idea that one might "take out" a Russian city if Soviet troops invade a country, and keep "taking out" one every day until they quit, has been occasionally adverted to journalistically but not systematically explored. Similar in spirit is the idea of hostile action on a small scale — sinking ships, blockading ports, jamming communications, or whatever it may be.

There are a number of Russian actions of an aggressive or hostile sort that might provide neither locale for a limited war nor the dramatic act to trigger massive retaliation: efforts to harass, blackmail, or blockade neutral countries or American allies, a peacetime campaign to jam our early-warning and other radar, tricks with nuclear weapons as part of a war of nerves, instigation of sabotage in NATO countries, flagrant support of insurrection, or even the use of unaccustomed violence in quelling disturbances within their own satellites. It may do little good to combat these actions by like measures of our own; it may also not be wise to insist that we are about to boil over into massive retaliation. If something were to be done, the deliberate creation of a small but appreciable shared risk of general war might be considered. (Or, if not, at least the purpose and significance of

---

[2] A recent serious discussion is Morton A. Kaplan, "The Strategy of Limited Retaliation" (Policy Memorandum 19 of the Center of International Studies; Princeton, April 9, 1959).

Soviet mischief may need to be interpreted as an effort to intimidate by the creation of a shared risk of general war.)

How do we interpret a dramatic act like, say, limited nuclear reprisal on enemy territory? As in limited war, there again may be two parts to the "cost" imposed on an enemy. One is a direct cost: casualties, destruction, humiliation, or whatever it may be. The other is the created risk of all-out war. Nobody quite knows what happens if one country explodes a nuclear weapon in an enemy country. If the action is recognized as an isolated act, limited in intent, not part of a massive attack nor of a sneak attack against the other's retaliatory capability, the victim may not see wisdom in unleashing all-out war in response to the pain and insult. But, even if he does not, he is likely to do something that in turn will have consequences that may ultimately reach a stage of all-out war. If the response is simply to strike back in like fashion, the process may taper off, or it may explode. So, even if each side prefers to act cautiously, failure to understand completely how each other reacts might bring about a dynamic process that ultimately explodes into all-out war.

The odds may still be against it. Here again we are dealing with an action that *may or may not* bring about general war, the final outcome *not* being under the complete control of the participants, the probability of all-out war being a matter of judgment. To mention these possibilities is not necessarily to propose them, but to indicate how they should be interpreted. The sanction they impose on the victim — one that the threatener shares with him — is the recognizable increase in the likelihood of total war.

RISKY BEHAVIOR AND "COMPELLENT" THREATS

There is typically a difference between a threat intended to make an adversary *do* something (or cease doing something) and a threat intended to keep him from starting something. The distinction is in the timing, in who has to make the first move, in whose initiative is put to the test. To deter by threat an enemy's advance it may be enough to burn the bridges behind me as I face the enemy; to compel by threat an enemy's retreat I have to be committed to move forward, and this requires setting fire to the grass behind me with the wind blowing toward the enemy. I can block your car in the road by placing my car in your way; my deterrent threat is passive, the decision to collide is up to you. If you, however, find me in your way and threaten to collide unless I move, you enjoy no such advantage; the decision to collide is still yours, and I enjoy deterrence. You have to arrange to *have* to collide unless I move, and that is a degree more complicated.

The threat that compels rather than deters, therefore, often takes the form of administering the punishment *until* the other acts, rather than *if* he acts. This is so because often the only way to become physically committed to an action is to initiate it. Initiating steady pain, even if the threatener shares the pain, may make sense as a threat, especially if the threatener can initiate it irreversibly so that only the other's compliance can relieve the pain they both share. But irreversibly initiating certain disaster, if one shares it, is no good. Irreversibly initiating a moderate *risk* of mutual disaster, however, if the other's compliance is feasible within a short enough period to keep the cumulative risk within tolerable bounds, may be a means of scaling down the threat to where one is willing to set it going. Subjecting the enemy (and oneself) to a 1 per cent risk of enormous disaster for each week that he fails to comply is somewhat similar to subjecting him (and oneself) to a steady weekly damage rate equivalent to 1 per cent of disaster. (The words "somewhat" and "equivalent" may be interpreted very flexibly here.) [3]

"Rocking the boat" is a good example. If I say, "Row, or I'll tip the boat over and drown us both," you'll say you don't believe me. But if I rock the boat so that it *may* tip over, you'll be more impressed. If I can't administer pain short of death for the two of us, a "little bit" of death, in the form of a small probability that the boat will tip over, is a near equivalent. But, to make it work, I must really put the boat in jeopardy; just saying that I may turn us both over is unconvincing.

[3] To initiate risky action, if one cannot initiate it irreversibly, does not necessarily "win" over an opponent: the latter may still hope, by acting firm, to induce the initiator to back down. One still has to win the "war of nerves" if the adversary chooses to play it out for a while. But at least this symmetrical situation replaces one in which the asymmetry favored the opponent, who won by default if neither side acted.

Ideally, for this purpose, I should have a little black box that contains a roulette wheel and a device that will detonate in a way that unquestionably provokes total war. I then set this little box down, tell the Russians that I have set it going so that once a day the roulette wheel will spin with a given probability (numerically specified and known to the Russians) that, on any day, the little box will provoke total war. I tell them — *demonstrate to them* — that the little box will keep running until my demands have been complied with and that *there is nothing I can do to stop it.* Note that I do not insist that I shall *decide* on total war, or initiate it deliberately, if the box hits the critical combination. I leave it all up to the box which *automatically* engulfs us both in war if the right (wrong) combination comes up on any day.[4]

Given the fact that, even if the enemy complies, there is some risk that the box detonates war before he has a chance to collect himself and do our bidding, there is an advantage in making it less than certain that the box will explode on any given day. In ordinary deterrence — where nothing happens *unless* the enemy acts contrary to our demand — to threaten too much may be superfluous but not self-defeating; in the present case — where the threat starts fulfilling itself at a specified rate over time as soon as we commit ourselves to it — too big a threat can defeat its purpose. In this situation the small-probability threat is not just a possible substitute for the large certain threat; it is a superior and necessary alternative.

Take an example. A European country, having acquired a modest nuclear retaliatory force, tells the Russians to get out of Hungary or it will work terrible damage on the USSR. The Russians ignore the threat, since there is no persuasive way for the threatening country to make itself *have to* do anything so suicidal. Alternatively, the country threatens to send a missile a day over the USSR, with a nuclear weapon and a random device that explodes it somewhere over Russia if it hasn't been shot down. The Russians say they do not believe the country would do

---

[4] The tactic may be the less risky, the more automatic the mechanism is; the more automatic it is, the less incentive the enemy has to test my intentions in a war of nerves, prolonging the period of risk.

it; the country does it. The Russians protest and threaten, a day passes, the country does it again. Maybe one weapon gets through and detonates, maybe several do, maybe none do; if some do, maybe they burst over cities, maybe over populated countryside, maybe over deserted areas. The country keeps it up.

What is the country doing? The principal thing the country is doing — in addition to damaging or humiliating Russia — is incurring a painful risk that both it and Russia (and the rest of the world) will be engaged in all-out war in the near future, a war that neither it nor Russia wants. The country is saying in effect, "If you do not get out of Hungary, we *may* cause an all-out war to occur." By when must the Russians get out? The sooner they get out, the sooner the risk of war (from this cause) will be terminated or reduced. The country applying the pressure is not saying, "Get out or we shall deliberately start a war." The decision is not up to them, and does not depend on their displaying the manifest resolution for a final act. The Russians may suppose that the country concerned will do everything it can to prevent total war; but they also have to recognize that with these things flying around, exploding now and then, and with themselves responding in whatever way they feel obliged to, it is not altogether clear that the country concerned, and the Russians, know how to keep total war from occurring.

This illustration is intended just as an analogy for other actions in which posing a risk of all-out war may not be so recognizable as an integral part of what is happening. To take a more immediate situation, suppose an armored column were sent to Berlin in the event that ground access were denied, or suppose, once a transport squeeze on Berlin became intolerable, troops were sent in to claim and hold a corridor; suppose actions were taken that, whether intended to or not, generated some likelihood of an East German uprising. How do we analyze the nature of the pressure on the Russians? I think the answer is in large part that they are confronted with a risk of a war that both sides badly want not to occur, but that both sides may not be able to prevent. A rationale for direct action, even on a scale that by itself might accomplish little, could be the deliberate creation of a risk that we share with the Russians, providing them with the option either to terminate the risk by acting or to withdraw to meet our objectives.

This is not the only interpretation of such action, of course. It may be that we could win militarily if the fight stays on a small scale, and that for the Russians to enlarge it would require a discontinuous jump that they would be deterred from taking for fear of provoking a discontinuous response. In that case the initial limited war would contain a "deterrent" threat against enlargement of the war. Even so, an important reason why the threat of even small-scale war might be effective is that such a war promises a small but appreciable increase in the probability of an enormous war, the probability being small enough that the Russians believe the West could bring itself to create it, large enough to make it unprofitable for them to let it occur.[5]

It is worth noting that this interpretation suggests that the threat of limited war may be potent even when there is little expectation that we would win it. In these terms, a limited local war is not just local military action; it contains an element of "retaliation" on the Soviet homeland — not a small *bit* of retaliation, but a small *probability* of a massive war.

BRINKMANSHIP

The argument of this paper leads to a definition of brinkmanship and a concept of the "brink of war." The brink is not, in this view, the sharp edge of a cliff where one can stand firmly, look down, and decide whether or not to plunge. The brink is a curved slope that one can stand on with some risk of slipping, the slope gets steeper and the risk of slipping greater as one moves toward the chasm. But the slope and the risk of slipping are rather irregular; neither the person standing there nor onlookers can be quite sure just how great the risk is, or how much it increases when one takes a few more steps downward. One does not, in

---

[5] In the author's opinion the dispatch of United States troops to Lebanon in 1958 was not only both risky and successful but successful precisely because of the risk — a risk that the Communists could lessen or aggravate according to their response.

brinkmanship, frighten the adversary who is roped to him by getting so close to the edge that if one *decides* to jump one can do so before anyone can stop him. Brinkmanship involves getting onto the slope where one may fall in spite of his own best efforts to save himself, dragging his adversary with him.[6]

Brinkmanship is thus the deliberate creation of a recognizable risk of war, a risk that one does not completely control. It is the tactic of deliberately letting the situation get somewhat out of hand, just because its being out of hand may be intolerable to the other party and force his accommodation. It means harassing and intimidating an adversary by exposing him to a shared risk, or deterring him by showing that if he makes a contrary move he may disturb us so that we slip over the brink whether we want to or not, carrying him with us.

The idea that we should "keep the enemy guessing" about our response, particularly about *whether* we shall respond, needs an interpretation along these lines. It is sometimes argued that we need not threaten the enemy with the certainty of retaliation or the certainty of resistance, but just scare him with the possibility that we may strike back. This idea may be misconceived if it means confronting the Russians with a possible response that remains for us to decide on, one way or the other. The Russians may guess that after the event we should prefer not to strike back, particularly if they perform their aggression in moderate bites; and if we are unwilling to arrange so that we *have* to strike back, and are even unwilling to *say* that we certainly shall, we may seem to confirm their understanding of what our preference would be if we left ourselves any escape. So, if we are afraid that an absolute commitment to the threat might fail in its purpose and commit us to an action we prefer not to be committed to, there may be little to salvage by trying to persuade the enemy that we just might decide to do it anyway.

But the situation is different if we get into a position where it is clear to the Russians that we are sufficiently involved that, while we probably have a way out, we *may* not. To say that we may or may not retaliate for an invasion of some neutral country, depending on how it suits as at the time, and that we shall not

[6] Children understand this perfectly.

let the enemy make this decision for us, nor let him know just what to expect, may confront the enemy with what appears to be a bluff. But to get so involved in or near a neutral country with troops or other commitments that we are not altogether sure ourselves about whether we could evade a fight in case of invasion, may genuinely keep the enemy guessing.

In sum, it may make sense to try to keep the enemy guessing as long as we are not trying to keep him guessing about our own motivation. If the outcome is partly determined by events and processes that are manifestly somewhat beyond our comprehension and control, we create *genuine* risk for him.

THE IMPERFECT PROCESS OF DECISION

Underlying this threat that one "may" retaliate or precipitate war — the decision being somewhat beyond his control — is the notion that some of the most momentous decisions of government are taken by a process that is not entirely predictable, not fully "under control," not altogether deliberate. It implies that a nation can get even into a major war somewhat inadvertently, by a decision process that might be called "imperfect" in the sense that the response to particular contingencies cannot exactly be foretold by any advance calculations, that the response to a particular contingency may depend on certain random or haphazard processes, or that there will be faulty information, faulty communication, misunderstanding, misuse of authority, panic, or human or mechanical failure.

This idea does not reflect an unusually cynical view of the decision process. In the first place, decisions do have to be taken on the basis of incomplete evidence and ambiguous warning; and it is unreasonable to deny *in principle* the possibility of an irrevocable action taken on a false alarm. (Furthermore, one need not be obsessed with the likelihood of false alarm to recognize that there may be levels below which this particular danger cannot be pushed without incurring other dangers that outweigh it!)

Second, war can occur because both sides become committed to irreconcilable positions from which neither is willing to back down, particularly if backing down requires assuming, even momentarily, a condition of military vulnerability. And it takes no cynic to recognize that two governments may misjudge each other's commitments.

But in the third place, even an orderly government with responsible, comparatively cool-headed leaders is necessarily an imperfect decision system, especially in crises. This is so for a number of reasons, one of which is that in anything but a completely centralized dictatorship a number of persons participate in a decision, and they do not have identical value systems, judgments of enemy intentions, and estimates of military capabilities. A decision taken quickly in crisis may depend on who is present, on whether particular studies have been completed, on the initiative and forcefulness shown by particular leaders and counsellors who are reacting to a quite unprecedented stimulus. Some parts of the decision may be taken on delegated authority, and the person to whom the decision is delegated cannot necessarily reproduce the decision that would have been reached by a president or premier or cabinet in consultation with congressional or parliamentary leaders. There may even be some necessary contradictions in the decision process, such as constitutional issues that cannot be settled in advance but that make it difficult to prepare fully for certain contingencies because the necessity to break law or precedent can be accepted only implicitly, not explicitly prepared for. Finally, the need to keep secrets puts limits on the amount of advance preparation for contingencies that can be carried out.

For this reason there is no such thing as a "firm" plan, intention, or policy of a government to cover every contingency — even all important foreseeable contingencies. How the considerations add up, what interests are brought to bear, and how the collective decision procedure works in future crises is simply not fully determinable in advance.

If on top of this we recognize that there are ordinary human limitations on the intellectual and emotional ability of governmental decision makers during the conduct of dangerous maneuvers on the brink of war, it ought to be clear that there is such a thing as getting into a situation from which it looks as though the nation may successfully extricate itself but in which there is some appreciable risk that, try as it does within the limits it allows itself, it may not succeed.

One does not expect a government to call attention to its own failings in this regard and to communicate to an enemy that this incomplete mastery of its own actions is an integral part of its strategy. There are also powerful public-relations reasons for not pointing out to an enemy that one is even slightly susceptible to disastrous errors in judgment and false alarms, or that one is a little unsure how to escape from a risky situation. It is understandable, too, that a government engaged in limited war does not state that it has been attracted to this military action by the possible risk of all-out war that it entails. The point is that these things go without saying.

But the basic idea of a threat that leaves something to chance is important even if we do not consciously use it ourselves, even tacitly. In the first place it may be used against us. In the second place, we may misjudge some of the tactics we do use if we fail to recognize the presence of a risk-of-total-war ingredient that may be a significant part of our influence on the enemy even if we have never appreciated it. If — to take an example — this is an important part of the role of limited-war forces in Europe, our analysis of that role may be seriously mistaken if we do not recognize it. The usual idea that a trip wire either does work or does not work, that the Russians either expect it to work or expect it not to work, is mistaking two simple extremes for a more complicated range of probabilities.

PART IV

SURPRISE ATTACK:
A STUDY IN MUTUAL DISTRUST

9

THE RECIPROCAL FEAR OF
SURPRISE ATTACK

If I go downstairs to investigate a noise at night, with a gun in my hand, and find myself face to face with a burglar who has a gun in his hand, there is danger of an outcome that neither of us desires. Even if he prefers just to leave quietly, and I wish him to, there is danger that he may *think* I want to shoot, and shoot first. Worse, there is danger that he may think that *I* think *he* wants to shoot. Or he may think that *I* think *he* thinks *I* want to shoot. And so on. "Self-defense" is ambiguous, when one is only trying to preclude being shot in self-defense.

This is the problem of surprise attack. If surprise carries an advantage, it is worth while to avert it by striking first. Fear that the other may be about to strike in the mistaken belief that we are about to strike gives us a motive for striking, and so justifies the other's motive. But, if the gains from even successful surprise are less desired than no war at all, there is no "fundamental" basis for an attack by either side. Nevertheless, it looks as though a modest temptation on each side to sneak in a first blow — a temptation too small by itself to motivate an attack — might become compounded through a process of interacting expectations, with additional motive for attack being produced by successive cycles of "He thinks we think he thinks we think . . . he thinks we think he'll attack; so he thinks we shall; so he will; so we must."

It is interesting that this problem, though it arises most dramatically in situations that would usually be characterized as conflict, like that between the Russians and us or between the burglar and me, is logically equivalent to the problem of two or more partners who lack confidence in each other. If each is under some temptation to abscond with the joint assets; if each has a little suspicion that the other may be contemplating the same thing; if each realizes that the other may suspect too, and may suspect himself the object of suspicion; we have a pay-off matrix identical with that of a surprise attack problem. If the heat is on some members of the mob, the rest of the mob may be tempted to rub them out to keep them from squealing, and those in danger may be tempted to squeal in self-defense. So the game structure of "preclusive self-defense" is the same as that of "partnership confidence."

The intuitive idea that initial probabilities of surprise attack become larger — may generate a "multiplier" effect — as a result of this compounding of each person's fear of what the other fears, is what I want to analyze in this chapter. More particularly, I want to analyze whether and how this phenomenon can arise through a *rational* calculation of probabilities or a *rational* choice of strategy, by two players who appreciate the nature of their predicament. The intuitive idea itself, even if misconceived, may be a real phenomenon and motivate behavior; people may vaguely think they perceive that the situation is inherently explosive, and respond by exploding. But what I want to explore is whether this phenomenon of "compound expectations" can be represented as a rational process of decision. Can we build an explicit model of this predicament in which two rational players are victims of the logic that governs their expectations of each other? [1]

INFINITE SERIES OF PROBABILITIES

We might begin by trying to set up the problem as follows. A player operates on a set of probabilities, a potentially infinite series of them. First is the estimated probability, P_1, that the other party "really" prefers to attack, that is, that the other will attack even if he does not fear an attack himself. Second is the probability, P_2, that the other player *thinks* that I "really" prefer to attack him, that is, that I will attack him even if I do

[1] Game theorists will recognize this problem as the nonzero-sum counterpart to what, for zero-sum games, has been called a "dueling game." The nonzero-sum version considered here involves the question of whether to shoot, not when to shoot.

not fear an attack on me. Third is the probability, P_3, that *he* thinks *I* think *he* "really" would; fourth is the probability, P_4, that *he* thinks *I* think *he* thinks *I* "really" would. Fifth, sixth, seventh, and so on are built up by lengthening the train of "he thinks" and "I think" with a separate probability attached to each member of the series. The over-all probability that he will attack is then given by:

$$1 - (1 - P_1)(1 - P_2)(1 - P_3) \cdots .$$

The trouble with this formulation is that nothing generates the series. Each probability is an *ad hoc* estimate, reflecting additional data about the specific information structure of the particular situation. We cannot, starting with a few terms in the series as data, project the rest to infinity, or however far it goes, and operate mathematically on the whole series. The number of terms in such a series can be only as much as a player has time to estimate, or the intellectual stamina to keep in mind, since he has to produce each new term of the series by an independent estimating process. It is true, we might set up particular games with information structures that would yield a formula for the series — for example, a series of spins of a roulette wheel determine whether the other player is told my "true" value system, whether I am told whether he has been told, whether he is told whether I have been told what he was told, and so forth — but these would be special games, and might not illuminate much the general situation we are trying to come to grips with. What we need is a formulation of the problem that permits us to work with a limited number of arbitrary parameters, representing perhaps the initial or "objective" terms in a series, in a context that automatically generates the values of any additional probabilities that may be conceived of through the indefinite reiteration of "He thinks I think." We need to formulate the problem in a way that makes each person's expectations a function of the other's.

A "STRICTLY SOLUBLE" NONCOOPERATIVE GAME

As a first try, we can assign to each of the two players a basic parameter representing the likelihood *that he would attack if he should not.* The values of these parameters are to be fully known and known to be known by both players. What I mean by "should not" is contained in the following two-part behavior hypothesis.

The first part of our behavior hypothesis is that, if the two players both perceive that a joint policy of no-attack is the best of all possible outcomes for both of them, they will recognize this "solution" and elect to abstain. If, for example, the pay-off matrix is as shown in Fig. 19, each will have confidence in their mutual

	I	II
i	0	−.5
	0	.5
ii	.5	1
	−.5	1

FIG. 19

confidence and will elect the strategy that yields both players the best possible outcome. This seems to be a fairly modest demand on the rationality of the two players.[2] (It is a questionable one, I suppose, mainly if the superiority of joint no-attack over unilateral surprise attack is small, too small to make both players completely confident that they understand each other. And this possibility — that somebody will be tempted to break discipline just to be on the safe side, or for fear that the other may try to be on the safe side — is allowed for in the second part of the behavior hypothesis, immediately following.)

The second part of the hypothesis is that there is some probability, P_r, for player R, and P_c for player C, that the player will in fact attack when he elects (or should elect) a strategy of no-attack, that is, that his decision will contradict the first part of our hypothesis. This is what was meant by the notion that a player might attack even when he "should not." Just what this

---

[2] In the terminology of Luce and Raiffa, if the noncooperative game has a "solution in the strict sense," that "solution" is here assumed to prevail. *Games and Decisions*, p. 107. Actually the condition is somewhat stronger here, since the solution is jointly preferred by the two players over all alternative outcomes, not just over all other equilibrium points.

parameter represents we shall leave open: it may be taken to be the probability that the player is irrational, or the probability that the pay-off matrix is misconceived and that he "really" prefers unilateral surprise attack, or the probability that somebody will make a mistake and inadvertently send off the attacking force. This parameter, for each player, is "exogenous" in our decision model: it is a datum provided from outside. It is not generated by the interaction of the two players.

These two parameters, P_c and P_r, are assumed to be plainly visible to the two players; there is nothing secret or conjectural about them. This assumption might seem to beg the question we are trying to answer, but it does not. These two exogenous likelihoods of attack do not by themselves indicate what the probability is that the players will in fact attack. They are only one element. The problem is to see whether, given these basic sources of uncertainty, the interaction of the two players' expectations generate additional motive to attack. We have to put at least *some data* into the problem for expectations and conjectures to work on. The only way to hold the arbitrary inputs to a minimal level is to make these two parameters fully visible; otherwise we must state what each guesses about them, what he guesses the other to guess about them, what he guesses the other to guess that he himself guesses about them, and so on. Again we would have the infinite series of *ad hoc* specifications, with the extra difficulty of dealing with probability distributions of probability distributions. The only way to break clean, and to provide a point of departure for calculating what each should fear the other to fear, is to make this one basic uncertainty for each player a matter of record. What we want to see is how an "objective" source of basic uncertainty generates a superstructure of subjective anxieties about each other's anxiety.

We now have a situation that looks as though it would generate the compound self-defense situation that we spoke of. The first player must consider whether the other player's likelihood of attack is serious; he must also consider that the other player is reciprocally worried. Even a player whose own probability of "irrational" attack is known to be zero must consider that the second may attack not only irrationally but also out of fear that the first, fearing the second's attack, may try to strike first to forestall it. Thus it does seem as though we might get a compounding of motives.

But we do not. We do not get any regular kind of "multiplier" effect out of this. The probabilities of attack by the two sides do not interact to yield a higher probability, except when they yield certainty. That is, the outcome of this game, starting with finite probabilities of "irrational" attack on both sides, is not an enlargement of those probabilities by the fear of surprise attack; it is either joint attack or no attack. That is, it is a pair of *decisions*, not a pair of *probabilities* about behavior.

We work this problem by recomputing the pay-offs in the original matrix, using the two parameters representing the probability of "irrational" attack. The upper left cell in the matrix stays as it was. The lower right cell has its pay-offs recomputed, as a weighted average of the four cells. For, if both players choose the strategy of no-attack, there is a probability equal to $(1 - P_c)(1 - P_r)$ that no attack will occur, a probability equal to $P_r(1 - P_c)$ that R will attack and C will not, a probability equal to $P_c(1 - P_r)$ that C will attack and R will not, and a probability equal to $P_c P_r$ that both will attack. In the same way, the pay-offs in the lower left cell are a weighted average of the pay-offs in the lower row; for if C elects to attack, he certainly does attack, while if R *elects* not to, he actually does or does not with probabilities P_r and $(1 - P_r)$ respectively. Thus with probabilities of irrational attack equal to 0.2 for each player, our original matrix would yield a modified matrix like the one in Fig. 20.[3] With proba-

I   II

	0	−.4
i	0	.4

	.4	.64
ii	−.4	.64

FIG. 20

[3] In effect we view the players as choosing — in the language of game theory — between one "pure" strategy and one "mixed" strategy the mixture specified by an autonomous parameter. (They could, of course, further mix the pure and mixed strategies, but in the present instance there is no reason to.)

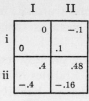

FIG. 21

bilities of irrational attack equal to 0.7 for C and 0.2 for R, we get Fig. 21. And with probabilities of 0.8 apiece for irrational attack, we get Fig. 22.

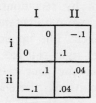

FIG. 22

The probabilities of irrational attack in the first of our modified matrices, namely the probabilities of 0.2 for each of the players, prove to be innocuous. That is, they are innocuous *with respect to the choice of strategies.* They yield a new pay-off matrix that still has a "strict solution" in the lower right corner. The *value of the game is reduced* for each player, since there is no escaping those two basic probabilities; but the *contemplation* of the probabilities has not led to their aggravation. Each player has fully taken them into account, has seen that there is still a jointly preferred solution at no-attack, and by the original hypothesis has chosen that strategy.

The last of our modified matrices, with a 0.8 probability for each player, is symmetrical and unstable; each player would now rather attack than hope for joint no-attack, and each knows that the other would too. This is a perverse situation, corresponding to the "prisoner's dilemma" familiar in game theory; the only efficient solution would be a binding agreement to elect no-attack (which still leaves them suffering the reduced value

214 SURPRISE ATTACK

of o.o4), if binding agreements were institutionally possible and if play were forcibly postponed to give the players a chance to reach such an agreement.[4]

The second of the modified matrices is also unstable, though not in a symmetrical way. Player C's likely irrationality requires player R to anticipate it by attacking in self-defense; player C, knowing this, attacks too.[5]

[4] "Prisoner's dilemma" refers, in game theory, to a configuration of payoffs that gives both players dominant incentives — in the absence of an enforceable agreement to the contrary — to choose strategies that together yield both players a less desirable outcome than if both had made opposite choices. The name derives from the problem of two prisoners, separately interrogated, who may confess to a moderate crime in common or accuse each other of a heavy crime, an accuser going free unless himself accused, the accused one or ones receiving heavy sentences. See Luce and Raiffa, pp. 94 ff.

[5] A somewhat different, and rather interesting, case occurs if we put P_r equal to o.2 and P_c equal to o.6. The modified matrix (for R only) is then:

P_r

0	.2
−.4	.12

R still has a "dominant strategy" of attack; he does better by attacking, no matter what C does. But in this case, as distinct from the case portrayed in Fig. 19, he is worse off than if neither side had elected to attack. It is C's knowledge of R's dominant strategy that causes them both to get zero. C's "irrationality," expressed in P_c, provides R with a motive for attacking in "self-defense"; but an element in that motive — a small "impurity" in the self-defense motive — is R's possibility of achieving surprise and thus of doing better than just meeting an incoming attack. If R were incapable of surprising C, even when he tried, his pay-off in the upper right cell of the original matrix would be zero, not o.5, and the modified matrix for R would be:

0	0
−.4	.08

This "worsens" both pay-offs for R in the right-hand column, but the upper more than the lower. It therefore eliminates R's motive to attack, and C knows it, so the outcome is at joint no-attack. Not only, then, may it help both players if the more "irrational" member is incapable of attack; it may even help them both if the "victim" is incapable of achieving surprise even in

The limits to the values of our two parameters, P_r and P_c, beyond which they make the situation unstable and provoke joint attack, are — letting h stand for the value obtained by unilateral surprise attack, $-h$ the value obtained by being attacked while not attacking, o the value obtained by simultaneous attack, and 1.o the value of joint no-attack, for each player —

$$P_c < 1 - h_r,$$
$$P_r < 1 - h_c.[6]$$

Figure 23 illustrates what happens to the "value of the game" for each player, and for each strategy, as one of the P's varies from o to 1.o. Putting P_r equal to o.2, and plotting the values of the game against P_c (based on the matrix of Fig. 19), yields values for C and R as diagrammed. At $P_c = 0.5$, the game be-

---

"self-defense." The condition for this special case, in terms of the parameters used in the next paragraph in the text, is

$$1 - h < P_c < 1/(1 + h).$$

This point can be made more general. Suppose the value of "winning" a war, denoted by h, may exceed 1; if it does, and if it is always a winning strategy to attack when the other does not, both players have dominant strategies at "attack." They both gain zero, when they might have had more if they could have abstained. Suppose, now, that the probability of achieving surprise, and thereby winning, is only Q, so that the expected value to be achieved through unilateral attack is only Qh. If Qh is less than 1, we are back to a matrix with a strictly preferred solution at joint no-attack; and, allowing for the probability of "irrational" attack, the game is stable if $P_c < 1 - Q_c h$ and $P_r < 1 - Q_r h$. Suppose that P_c and Q_c meet the first of these conditions: then it is to R's advantage, as well as C's, that the second condition also be met. If P_r is beyond manipulation, R should wish that Q_r, his own capacity for surprising an enemy, should be less than $(1 - P_r)/h$. Only then can he, and C, gain more than zero. If R can, at his own expense, improve his "enemy's" alert system, or if he can blunt his own surprise capacity in a visible way, to hold Q_r below the limit, he should do so. The principle is the same as that of two partners, somewhat distrustful, who keep two separate private padlocks on the partnership vault. If one could not afford a padlock, the other should provide it to him at his own expense; only then can they do business together.

[6] A more general formula, covering the nonsymmetrical case, and using R_{11}, R_{12}, R_{21}, R_{22} to denote the pay-offs to R in row 1 col 1, row 1 col 2, and so on,

$$\frac{P_c}{1 - P_c} < \frac{R_{22} - R_{12}}{R_{11} - R_{21}}$$

The numerator is the "cost" of erroneously attacking; the denominator is the "cost" of erroneously failing to attack. The criterion is the same, it may be noted, as if P and $(1 - P)$ were sure probabilities rather than probabilities of departure from, and adherence to, a "rational" behavior pattern.

comes unstable, and the value of the game goes to 0 for both players.

That this game does not quite correspond to the original notion of "compounded probabilities" is exemplified by the fact that we

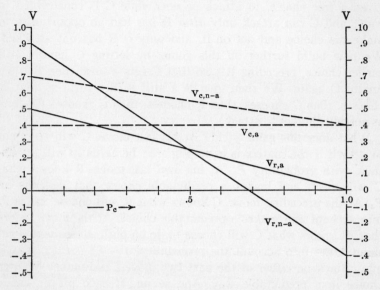

FIG. 23. Value of the game to R and C, as a function of P_c; $P_r = 0.2$. $V_{r,\ n-a}[= 0.9 - 1.3P_c]$: value of game to R, joint strategy of no-attack; $V_{r,\ a}[= 0.5 - 0.5P_c]$: value of game to R, who attacks while C elects not to; $V_{c,\ n-a}[=0.7 - 0.3P_c]$: value of game to C, joint strategy of no-attack; $V_{c,\ a}[= 0.4]$: value of game to C, who attacks while R elects not to.

can ignore the lesser of the two parameters if they are unequal. If both are below the critical limit, it does not matter what they are; if one is over the limit by ever so little, it makes no difference whether the other is 0 or 1.0. They can thus be potent beyond what they do to the value of joint nonattack, because they can cause the players to shift from a strategy of no-attack to a strategy of attack. But they do so in an all-or-none way. The *likelihood* of attack either is confined to the exogenous likelihoods, or becomes certainty.

THE GAME AS A SEQUENCE OF MOVES IN TURN

We get the same result if we try a game with *moves-in-turn* for the pay-off matrix that we have been using. Suppose R is given a free choice, to attack or not, while C is constrained to wait; and C can attack only *after* R has had an opportunity to make his choice and act on it, and only *if* R has not attacked. We now build further on this game, by letting C have a still earlier choice, preceding R's, so that C gets a turn, then R, then finally C again. We then give R a still earlier turn, so that R chooses, then C chooses, then R chooses, then C chooses (as long as nobody chooses to attack).

What does this game yield? At his last move, C will *elect* not to attack if the matrix is as in Fig. 19; he actually will attack, then, with probability P_c. At his own last move, R knows what C will elect, and makes a predictable choice that depends on P_c. At the preceding move, C knows what R will choose, takes P_r into account and makes a predictable choice. At the move before that, R knows what C will choose to do on both subsequent occasions, takes into account the probability, $1 - (1 - P_c)^2$, that C may attack on either of the next two moves, and makes his own choice in a predictable way. And so on. If each player has n moves, with probability P_r or P_c of irrational attack *at each move*, the outcome depends on whether $\bar{P}_c = 1 - (1 - P_c)^n$ and $\bar{P}_r = 1 - (1 - P_r)^{n-1}$ meet the conditions derived earlier. If so, each player knows that the other will not subsequently choose to attack, and himself chooses not to at all turns. But if $\bar{P}$ exceeds the limit for one of them he will prefer to attack and the other knows it, so whoever has first turn attacks at once.

In other words, we are compounding probabilities, but still with an all-or-none effect, and without either player having to *combine* both players' irrationality parameters in the compounding process. Either the probability for at least one of them is big enough and the game long enough to cause the first player to attack, or else no one attacks. And, if we make the over-all probabilities of irrational attack independent of the number of turns, by letting the probability at each turn be equal to

$1 - (1 - P)^{1/n}$, so that the compounded total is just P_c or P_r, the outcome of this game is *independent of the number of turns*. If we think of this game, then, as an *analogy* of the he-thinks-I-think situation, with each turn symbolizing a cycle in the spiral of suspicions, we have a model in which the successive reciprocal fears of what each other fears make no difference: either there is "objective" basis for one of the players to attack, or they abstain.

RECONSIDERATION OF THE PROBLEM

The same seems to be true now if we go back to that burglar downstairs. If he behaves "rationally" as defined in our behavior hypothesis above, he must consider the likelihood that I will shoot him out of sheer preference; and he must consider that I may shoot him if I think there is a strong likelihood that he will shoot me out of sheer preference. But, if we both know what these two basic (exogenous) "likelihoods" are, we need not go any further. Either these basic probabilities are sufficient to make at least one of us shoot to forestall surprise, and hence to make both shoot, so that the second and higher degree fears are superfluous, or else they are insufficient by themselves to make either of us shoot in self-defense, and we know it and have nothing to fear beyond the exogenous likelihoods themselves. If we both can plainly see that neither would be quite induced to shoot solely out of fear of the exogenous probability that the other "really" wants to shoot, then we ought to be able to see that neither needs to fear preclusive action, that neither then needs to fear that the other fears it, and so on.[7]

But a different situation obtains if I shoot not by calculation but by nervousness. Suppose that my nervousness depends on how frightened I am, and my fright depends on how likely I think it that he may shoot me; and suppose he acts the same way. Then when I consider the exogenous probability that he may shoot me out of sheer preference, it makes me nervous; this

---

[7] For example, if the two could just communicate and check each other's understanding, they could reach an informal agreement not to *elect* to shoot that would leave no incentive to cheat — assuming, still, that the two basic parameters are clearly evident to both of them.

nervousness enhances the likelihood that I may shoot him even though I prefer not to. He sees my nervousness and gets nervous himself; that scares me more, and I am even more likely to shoot. He sees this increment in my nervousness, and matches it with one of his own, scaring me further; and the probability that I will shoot goes up again. Now we can denote each person's nervousness as a function of the other's, and the likelihood of shooting as a function of nervousness, and have a simple pair of simultaneous differential equations that seem to yield precisely the kind of phenomenon we started off to study.[8]

And the reason they do is that this model does not involve *criteria for decision*; that is, it does not involve a behavior hypothesis that tells us which of two strategies a person will select. Instead, our "nervousness model" is one in which people respond to the fear of attack by *a change in the likelihood* that they will themselves attack. Only in this way, by dealing with the *probability* of a player's decision, and not with a *rule* for decision — that is, not with a model in which the player calculates his best strategy and follows it — can we get the kind of "mutual aggravation" phenomenon that I described at the beginning of this chapter.

Now, does this mean that our phenomenon is not one that can be displayed by rational, decisive players? How can we envisage a player reacting to a change in his environment, or to a new bit of information, by *deciding* that he will do something "somewhat more probably" than before? A rational man may be nervous, in which case our theory is physiological rather than intellectual; but can we conceive of the rational game player's taking another look at the burglar and changing the adjustment on his roulette wheel?[9]

[8] There is an important asymmetry in the problem as formulated here. We have allowed for the possibility that one may shoot when he shouldn't and the other knows it — the "nervousness" case — but not for the possibility that one may *not* shoot when he ought to, and the other knows it. (There may be some chance that the burglar has wet ammunition or forgot to load his gun, and I may know that there is such a chance, he may know that I know it, and so forth.) This possibility would apparently be stabilizing, tending to reduce the likelihood of a *decision* to attack as well as the exogenous likelihood of inadvertent or irrational attack.

[9] Note that the usual rationale for a mixed strategy — that is, for rationally

Of course, individual and group decisions may be different in this regard. We could think of a collective decision by vote, with different members having different value systems and hence different thresholds of reaction to the probability of being attacked, so that the size of a vote to attack would be a function of the estimated likelihood of being attacked. If the vote also depends heavily on chance factors, such as absentees on voting day, the *probability* of the required majority in favor of attack becomes a rising function of the probability of the enemy's own decision, which in turn is a function of the first collective player's probability. So we can get the phenomenon we want for "rational" players if we deem rational a collective player that has divergent values and a voting system.

There is, however, a way to adapt our model even to the single, decisive, rational game player. It may be of fairly wide generality in partnership and surprise-attack problems. And it directly involves a significant part of the actual problem of military surprise, namely the dependence of decision on an imperfect warning system, and the possibility of both "type-1" and "type-2" errors in the decision process.

PROBABILITY-BEHAVIOR GENERATED BY AN IMPERFECT WARNING SYSTEM

Presumably the danger of suffering a surprise attack can be reduced by the use of a warning system. But the warning system is not infallible. A warning system may err in either way: it may cause us to identify an attacking plane as a seagull, and do nothing, or it may cause us to identify a seagull as an attacking plane, and provoke our inadvertent attack on the enemy. Both possibilities of error can presumably be reduced by spending more money and ingenuity on the system. But, for a given expenditure, it is generally true of decision criteria that a tightening of the criteria with respect of one kind of error loosens them with respect to the other. To require less evidence of incoming

---

readjusting one's roulette wheel for decision — has no relation to the present case.

attack before "retaliating" is to require more evidence that they are really seagulls for holding back our own planes.

But now we can have a model of a rational decider who responds to an estimate of the probability of being attacked *not* by an overt *decision* to act or abstain, but *by adjusting the likelihood that he may mistakenly attack*. One's response to an increase in the probability of being attacked is to shift the criteria for decision that are used in the warning system in the direction of lesser likelihood of a failure to respond, and hence in the direction of greater likelihood of a false alarm that provokes one's own "retaliation." If each player's response to an increased danger of surprise attack is to enhance his own proclivity toward inadvertent attack, the *probability* of each player's attack is now a rising function of the other's.[10] Such a warning system is the rational, mechanical counterpart of our nervousness in facing the burglar.

To build such a model (symmetrically, for simplicity) we can again let h denote the value of "winning" a war, $-h$ that of "losing" a war, o the expected value of simultaneous attack (50-50 chance of winning or losing), and 1.0 the value of no war at all. (This time we can let h exceed 1, as long as $(1 - R)h$ in the matrix below remains below 1. But if "winning" gains a Pyrrhic victory, h will be a small fraction.) We assume that successful surprise wins the war; "successful surprise" means that one attacks when the other does not *and* that the other's warning system fails him. Let R denote the reliability of a player's warning system, that is, the probability that an attack, if it comes, will be identified and surprise forestalled. Then the pay-off matrix is as in Fig. 24.

The probability that a player will attack when he should not, that is, that he will when his rational choice "should" be against attacking (in the sense used earlier), will consist of two parts. One, denoted by A, is the exogeneous likelihood of irrational attack; it excludes the possibility of an attack provoked by false alarm. The probability of an attack through false alarm is de-

---

[10] As noted below, this is not *necessarily* so; if increased danger of being attacked is associated with reduced vulnerability of the enemy to surprise attack, it is possible for one's response to be in the direction opposite to that described in the text.

noted by B. Thus the two types of error in the warning system are represented by B and $(1 - R)$; and the main feature of the model is that $B = f(R)$, $f'(R) > 0$. That is, the more we reduce $(1 - R)$ as a source of error, the more we increase B, and vice versa.

	I	II
i	0	$-(1-R_c)h$
	0	$(1-R_c)h$
ii	$(1-R_r)h$	1
	$-(1-R_r)h$	1

Fig. 24

Each player's strategy choice concerns the pair of values for B and R that will minimize his expected losses, that is, maximize the expected value of the game for him. Letting V_r denote the expected value of the game for R, the warning-system problem for R is to choose the pair of values for R and B, consistent with $B = f(R)$, that maximizes [11]

$$V_r = (1 - P_c)(1 - P_r) + P_r(1 - P_c)h(1 - R_c)$$
$$- P_c(1 - P_r)h(1 - R_r)$$
$$= (1 - A_c)(1 - B_c)(1 - A_r)(1 - B_r)$$
$$+ (A_r + B_r - A_rB_r)(1 - A_c)(1 - B_c)h(1 - R_c)$$
$$- (A_c + B_c - A_cB_c)(1 - A_r)(1 - B_r)h(1 - R_r).$$

Additionally, pursuant to the earlier matrix analysis, R should examine the resulting "modified" pay-off matrix that results from using these "optimal" values of R_r and B_r, together with the observed (or expected optimal) values of R_c and B_c, to see whether joint no-attack is still the jointly preferred outcome. The conditions for a joint preference at no-attack, with optimally adjusted warning systems, would be:

[11] It is assumed for convenience of illustration that an inadvertent attack due to false alarm is the same kind of attack as a premeditated attack, with the same likelihood of achieving surprise. Also, we are ignoring the time dimension of B, which probably ought to be thought of as the probability of false alarm per unit of time, while $(1 - R)$ is the probability of error per incoming attack, and A might have some of both elements. Thus the time horizon is assumed fixed in this model.

$$P_o = (A_c + B_o - A_cB_o) < \frac{1 - h(1 - R_c)}{1 - h(R_r - R_c)},$$

$$P_r = (A_r + B_r - A_rB_r) < \frac{1 - h(1 - R_r)}{1 - h(R_c - R_r)}.$$

With symmetry, the denominators in the right-hand terms become just 1.

Actually, as will be seen below, this second examination may be unnecessary; for certain behavior hypotheses, "optimal" adjustment of R and B (for any value short of $R = 1$) requires that the conditions for stability of the modified matrix be met.

It remains to be specified how the players behave. Broadly speaking, we can make either of three hypotheses, corresponding more or less to the difference between "parametric behavior," a "tacit game," and a "bargaining game."

DYNAMIC ADJUSTMENT (PARAMETRIC BEHAVIOR)

First we may try supposing that each player takes the probability of being attacked as given, that is, as a parameter and not a variable in his own loss function, and does the same with the reliability of his opponent's warning system. That is, he directly *observes* the values of his opponent's B and R, and selects the pair of values for own B and R that minimize his expected losses. This assumption tends to make each person's choice of B a rising function of the probability that the other will attack. (It only "tends to," since there is a possibility that the corresponding change in the other's R provides an offsetting inducement, as mentioned below.) If we think of the two players as continually adjusting their values of B and R, each with an eye on the other's B and R, but always responding parametrically to the current probability of being attacked and not projecting the other's behavior as a function of his own, we get a simple dynamic "multiplier" system — stable or explosive depending on the parameter values and shape of the f function. We can express each player's optimum value of B as a function of the other's, solve the two equations, and deduce the stability conditions for the equilibrium. We can also compute "multipliers" relating each player's changes of B and R to shifts in the f function or to changes in the A parameters.

Explicitly, to find the "parametric-behavior" function for player R we maximize V_r with respect to R_r, subject to $B_r = f(R_r)$ but treating B_c and R_c as fixed. Using the formula given earlier for V_r, we get

$$f' = \frac{P_c h (1 - B_r)}{(1 - P_c) [1 - h(1 - R_c)] - P_c h(1 - R_r)}$$

and, for $h(1 - R_c) < 1 > h(1 - R_r), f'' > 0$.

Since f' is presumed positive, the denominator must be positive if V_r is to be maximized with $R < 1$; but the condition that the denominator be positive is precisely the condition that P_c must meet in order that player R still prefer joint no-attack. Thus, if both players have optimal adjustments with $R < 1$, those optimal values of R and B are also perforce consistent with joint preference at no-attack.

The relation of B_r to B_c under this behavior hypothesis, that is, the slope of the resulting function that yields R's optimal B-value for given values of B_c, is obtained by differentiating both sides of the above equation:

$$\frac{dB_r}{dB_c} \text{ (along player R's behavior function)} = \frac{dB_r}{dR_r} \frac{dR_r}{df'} \frac{df'}{dB_c}.$$

$$= \frac{f'}{f''} \frac{df'}{dB_c} = \frac{f'}{f''} \left(\frac{\partial f'}{\partial B_c} + \frac{\partial f'}{\partial R_c} \frac{dR_c}{dB_c} \right)$$

$$= \frac{f'}{f''} \left(\frac{\partial f'}{\partial B_c} + \frac{\partial f'/\partial R_c}{\phi'} \right),$$

where $B_c = \phi(R_c)$ denotes the corresponding function for player C.

Since $\partial f'/\partial R_c$ is negative, small values of ϕ' may make player R's dB_r/dB_c negative; it does so by raising the "cost" of inadvertent attack enough to outweigh the increase in the risk of being attacked. In other words, B_r is a function not just of B_c but of $\phi(B_c)$ as well; B_r tends to be increased for a rise in B_c but lowered for a rise in R_c, while B_c and R_c rise together as we consider moving out the B_c axis.

A stable equilibrium requires that player R's dB_r/dB_c and C's dB_c/dB_r should have a product less than 1, that is, that with B_r measured vertically and B_c horizontally, C's curve should intersect R's from below. The general "multiplier" expression relating changes in the B's and R's to shifts in the functions (or to changes in the values of the A's) contains 1 minus this product in the denominator.

As remarked earlier, the denominator in the expression for f' disappears, and R_r, B_r, and f' rise sharply, as h approaches the condition for an unstable matrix. (Actually, stability of the matrix game, as distinct from stability of a parametric-behavior equilibrium, is not a relevant concept for the parametric-behavior hypothesis; to contemplate the matrix and to anticipate the other's action is to project his behavior, not to observe it and adapt to it.)

It may also be noted that player R may ignore A_r in his calculations. It drops out of the formula for optimum B_r and R_r. Intuitively, this is because the only contingency in which *either* the value of R_r or the value of B_r can make any difference is the contingency that R *not* launch "irrational" attack; if he does, B and R are irrelevant to him. (However, A_r does affect the condition for a stable matrix, since it does enter into the condition that P_r must meet. So in projecting C's adjustment, R would have to take A_r into account. But "projecting" C's behavior, rather than just observing B_c and R_c continuously, would make R's behavior nonparametric, contradicting the present hypothesis. If player R were considering the value of spending money to improve his warning system, A_r would affect the calculation since it affects the probability that the system makes any difference; this consideration is outside the present model.)

A TACIT GAME

We can make another behavior hypothesis, which may lead to the same result. Instead of supposing that each player sees how the other's R and B are adjusted, takes them as given, and responds to them; we can suppose that each player knows the technological opportunities of the other player — the functional relation between R and B for the other player — but cannot reliably observe how the other has adjusted the values of R and B. That is, each understands the mechanics of the other's warning system, but can never be sure just what instructions the other has given on how to interpret the evidence that comes in over the system — the other's decision rule. This hypothesis yields us a noncooperative game, in which each player must choose a value for B (that is, for R), not knowing what value the other has chosen but knowing the other's pay-off matrix.

In this case, we have a pay-off matrix with an "equilibrium point" at precisely the point, if any, where the parametric-behavior hypothesis yielded a stable equilibrium.[12] In other words, what was the "solution" under the parametric-behavior hypothesis is still a candidate for being called a "solution" in the noncooperative form of the game. (In neither case is the equilibrium point necessarily unique. If it is not, the first hypothesis makes the outcome depend on initial conditions and "shocks"; the second tends to complicate the intellectual problem of identifying "solution" strategies.)

This solution, of course, is inefficient for the two players. It is an example of "prisoner's dilemma," mentioned above (p. 214); reciprocal increases in the values of the B's have simply raised the likelihood of attack by each side.[13] There are lesser values for the two B's that would make both parties better off; and if the probabilities of deliberate sneak attack by the two players are equal (A's are equal), an agreement to have no warning system at all, that is, no possibility of false alarm, would be the preferred bargain for the two parties if they were restricted to bargains that gave them identical warning systems.[14]

---

[12] An *equilibrium point*, in game theory, is a pair of strategies for the two players such that each is optimal vis-à-vis the other. (There may be several such points.)

[13] Economists may find the situation reminiscent of two producers who both allocate their limited productive resources between two commodities. One commodity, "security against false alarm," involves external economies; the other, "security against surprise," involves external diseconomies.

[14] If the A's, B's, and R's are equal, V_r and V_e are equal to $(1 - P)^2$, which has a maximum at $B = 0$. (If B has some minimum value greater than 0, we can attribute it to A.) If B's and R's are equal but the A's are not,

$$dV_r/dB = -2(1 - B)(1 - A_e)(1 - A_r) + (A_e - A_r)(h/f'),$$

A BARGAINING GAME

If we consider the possibility of the two players' negotiating to reduce the sensitivity of their alert systems, in the interest of mutual reductions in B at the cost of smaller R's, and assuming that enforcement of such an agreement were possible, there is no very convincing way of deriving a unique solution without further specification of the bargaining framework. If the *solution* has to be symmetrical and the *game* is symmetrical, that is, if they negotiate over a *common pair of values for R and B*, the result is as just mentioned — o values for B, even if this means o for R, no warning system at all. If warning systems are to be identical, there is some critical difference between the basic probabilities of deliberate sneak attack for the two people (between A_r and A_c) beyond which a side payment would be required for an agreement on the abolition of warning systems.

But, in general, this becomes a wide-open bargaining problem. It is even wider open than the present formulation suggests, since the players may not only manipulate values of R and B but of course can now threaten direct attack, or operate on the institutional arrangements that determine the values of the A's.

There is an enforcement difficulty with any agreement on reduction of values of R and B in the mutual interest; it is that each other's values of R and B may not be observable. They depend — at least to an important extent — on the criteria that will govern future decisions, not solely on the observable, physical mechanics of an alert system. They depend on how long one will wait to be "sure," and on what risks one will accept in an emergency. Furthermore, failure to keep the agreement, if it leads to anything, leads to war itself; so recriminations and damage suits are out of the question if our model represents all-out war

---

which can be positive with A_c greater than A_r and f' small. In this case one of the players — the one with smaller A — has a preference for *some* warning system even if it must be common to both of them, compared with none at all; but it involves lesser values for B and R than parametric behavior (or a noncooperative game) would lead to, as can be seen by putting the above expression equal to o and comparing the resulting formula for f' with that corresponding to parametric behavior.

rather than a border scrape or a minor transgression of one partner against another.

It might be that $R = B = 0$ is qualitatively observable — the physical "absence" of any system at all. Even this possibility is unavailable, as an enforceable system, if the matrix is unstable with $R = 0$, that is, with $h > 1$. In that case, some "risk" in the form of B is necessary to put the R's safely in the range where $h(1 - R)$ is less than 1.

It may also be difficult to have an agreement that explicitly recognizes the A's, since it may be politically difficult to admit that one's A is above 0.

The players may be driven then to rely on arrangements that either observably blunt their own capacity for surprise or observably improve their own and each other's transformation curves relating R to $(1 - B)$. Both sides may, for example, agree to spend more on the alert system, to make it more efficient; and the richer side may prefer to finance improvements in the other's alert system, rather than leave it in a form that either aggravates the other's sense of insecurity or makes him susceptible to false alarm. An agreement to design forces that have no surprise-attack potential, but instead have improved vulnerability to surprise attack themselves, would seem to be indicated. That is, instead of making R and B the terms of an agreement, they might be forced by the unobservability of R and B to work on the f and ϕ functions themselves, considering each of these functions to involve both one's own alert system and the enemy's (partner's) attack force. (It should be noted, however, that "innovations" in the warning systems — shifts of the f and ϕ functions in the direction of less B for a given level of R and vice versa — are not in all cases stabilizing. Those that raise the *marginal* cost of R may lead to higher values of B; these would be perverse innovations from the point of view of the two players together, analogous to an "improvement" in the prisoners' dilemma matrix that raises each player's pay-off for noncooperative strategies.)

The bargaining-game formulation also lends itself to bargaining-tactic analysis. For example, if one player acts parametrically and the other knows it and takes it into account, the first displays a "reaction function" [15] which goes into the other's formula for V which the latter tries to maximize. In general, the analysis of "strategic moves" of the kind discussed in Chapters 2, 5 and 7, are relevant to this version of the surprise-attack, partnership-discipline game.

MORE THAN TWO PLAYERS

An interesting variant of the problem would occur if the number of players were increased, or if a third player were brought in as an autonomous agent. To the extent that attack from other quarters must be anticipated, the incentive toward mutual reduction of alert systems is reduced. It remains true, however, that any two players in a larger game can find some advantage in jointly modifying their alert systems, in the direction of lesser danger of false alarm, by taking into account the "external diseconomies" for each other that they leave out of account when behaving parametrically. Two armed watchmen patrolling the same building, each subject to some temptation to shoot on sight, would be better off if they could find some way of reaching an enforceable agreement to be a little less ready to shoot on sight, to reduce the likelihood of shooting each other. (Actually, the two-watchmen problem is a representation of our original model, if we let our original parameters, P_c and P_r, represent the relative likelihoods that a man met in darkness is a burglar rather than the other watchman. We have to introduce some uncertainty about a burglar's behavior — that is, to let him join the game as a rational third participant trying to anticipate the others' decisions — in order to add complications to what we already had.) [16]

---

[15] Compare the note on p. 151 regarding the concept of "reaction function."

[16] Arthur Lee Burns, of the Australian National University, has discussed some interesting problems of a three-or-more person world. The deliberate provocation of war between two parties, by a mischievous third party, is a possibility when an overt act of ambiguous authorship can be introduced into the reciprocal-suspicion model; and the analysis takes on additional richness when one considers warning systems that, for technical reasons or by reason of joint custody, permit one or both of the central players to witness what is coming in on the other's radar screen. See his "Rationale of Catalytic War" (Center for International Studies, Research Memorandum No. 3; Princeton University, 1959).

10

SURPRISE ATTACK AND DISARMAMENT

"Disarmament" has covered a variety of schemes, some ingenious and some sentimental, for cooperation among potential enemies to reduce the likelihood of war or to reduce its scope and violence. Most proposals have taken as a premise that a reduction in the quantity and potency of weapons, particularly of "offensive" weapons and of weapons that either deliberately or incidentally cause great civilian agony and destruction, promotes this purpose. Some schemes have been comprehensive; others have sought to identify particular areas where the common interest is conspicuous, where the need for trust is minimal, and where a significant start might be made which, if successful, would be a first step toward more comprehensive disarmament. Among these less comprehensive schemes, measures to safeguard against surprise attack have, since the President's first "open-skies" proposal in 1955, come increasingly into prominence.

The focus on surprise attack has not reflected an abandonment of interest in a more ambitious dismantlement of arms; rather it represents the philosophy of picking an area where success is most likely, in order to establish some tradition of successful cooperation. The search for safeguards against surprise attack has generally been considered, in our government and elsewhere, not as an *alternative* to disarmament, but as a *type* of disarmament and a possible step toward more.

Nevertheless, though schemes to avert surprise attack may be in the tradition of disarmament, they represent something of an innovation. The original open-skies proposal was unorthodox in its basic idea that arms themselves are not provocative so long as they are clearly held in reserve — so long as their stance is deterrent rather than aggressive. The proposal was also unorthodox in its dramatic reminder that, important as it may be to keep secrets from an enemy and in some matters to keep him guessing about what our plans are, it can be even more important to see that the enemy is *not* left to speculate about our intentions toward surprise attack against him *if in fact we are not planning any such attack*. We are interested not only in assuring ourselves with our own eyes that he is not preparing an attack against us; we are interested as well in assuring *him* through *his* own eyes that *we* are preparing no deliberate attack against *him*.

The importance of not keeping that particular secret has an analogue in our alleged political inability to attack first. As General Leslie R. Groves remarked in a speech, "If Russia knows we won't attack first, the Kremlin will be very much less apt to attack us. . . . Our reluctance to strike first is a military disadvantage to us; but it is also, paradoxically, a factor in preventing a world conflict today." [1] We live in an era in which a potent incentive on either side — perhaps the main incentive — to initiate total war with a surprise attack is the fear of being a poor second for not going first. "Self-defense" becomes peculiarly compounded if we have to worry about his striking us to keep us from striking him to keep him from striking us. . . . The surprise-attack problem, when viewed as a problem of reciprocal suspicion and aggravated "self-defense," suggests that there are not only secrets we prefer not to keep, but military capabilities we might prefer not to have.

Of course, it is even better if the other side does not have them either. So there may be advantages in thinking of the surprise-attack problem as one suitable for negotiation.

The innovation in the surprise-attack approach goes further. It has to do with what the scheme is designed to protect and what armaments it takes for granted. An anti-surprise-attack scheme has as its purpose not just to make *attack* more difficult but to reduce or to eliminate the advantage of striking *first*. It must assume that if the advantage of striking first can be eliminated or severely reduced, the incentive to strike *at all* will be reduced.

[1] *The New York Times*, December 29, 1957, p. 20.

It is widely accepted that the United States has the military *power* virtually to obliterate the USSR, and vice versa. And it is widely accepted that, if either side struck the other a major nuclear blow, the nation so hit would have a powerful *incentive* to strike back with equal or greater force. But, if either side can obliterate the other, what does it matter who strikes first? The answer, of course, is that we are not particularly concerned with outliving the Russians by a day; we are worried about whether a surprise attack might have such prospects of *destroying the power to retaliate* as to be undeterred itself by the threat of retaliation. It is not our existing capacity to destroy Russia that deters a Russian attack against us, but our capacity to retaliate after being attacked ourselves. We must assume that a Russian first-strike, if it came, would be aimed at the very power that we rely upon for retaliation.

There is a difference between a balance of terror in which *either* side can obliterate the other and one in which *both* sides can do it no matter who strikes first. It is not the "balance" — the sheer equality or symmetry in the situation — that constitutes mutual deterrence; it is the *stability* of the balance. The balance is stable only when neither, in striking first, can destroy the other's ability to strike back.

The difference between a stable and an unstable balance is illustrated by another offensive weapon against which no good defense was ever devised.[2] The "equalizer" of the Old West made it possible for *either* man to kill the other; it did not assure that *both* would be killed. The tense consequences of this weapon system can be seen on TV almost any night. The advantage of shooting first aggravates any incentive to shoot. As the survivor might put it, "He was about to kill me in self-defense, so I had to kill him in self-defense." Or, "He, thinking I was about to kill him in self-defense, was about to kill me in self-defense, so I had to kill him in self-defense." But if both were assured of living long

---

[2] A military historian, commenting on the alleged "historical truth" that there has never yet been a weapon against which man has been unable to devise a counterweapon or a defense, reminds us that "after five centuries of the use of hand arms with fire-propelled missiles . . . no adequate answer has yet been found for the bullet" (Bernard Brodie, *The Absolute Weapon* [New York, 1946], pp. 30–31).

enough to shoot back with unimpaired aim, there would be no advantage in jumping the gun and little reason to fear that the other would try it. The special significance of surprise attack thus lies in the possible vulnerability of retaliatory forces. If these forces were themselves invulnerable — if each side were confident that its own forces could survive an attack, but also that it could not destroy the other's power to strike back — there would be no powerful temptation to strike first. And there would be less need to react quickly to what might prove to be a false alarm.

Thus schemes to avert surprise attack have as their most immediate objective the safety of weapons rather than the safety of people. Surprise-attack schemes, in contrast to other types of disarmament proposals, are based on *deterrence* as the fundamental protection against attack. They seek to perfect and to stabilize mutual deterrence — to enhance the integrity of particular weapon systems. And it is precisely the weapons most destructive of people that an anti-surprise-attack scheme seeks to preserve — the weapons of retaliation, the weapons whose mission is to punish rather than to fight, to hurt the enemy afterwards, not to disarm him beforehand. A weapon that can hurt only *people*, and cannot possibly damage the other side's striking force, is profoundly defensive: it provides its possessor no incentive to strike first. It is the weapon that is designed or deployed to destroy "military" targets — to seek out the enemy's missiles and bombers — that *can* exploit the advantage of striking first and consequently provide a temptation to do so.

In identifying the surprise-attack problem as the possible vulnerability of each side's retaliatory forces to surprise, we are at the point where measures against surprise attack differ drastically from more conventional notions of disarmament. We are also at the source of a number of anomalies and paradoxes that have to be faced if we are to recognize the virtues and defects of particular schemes and to comprehend the motives behind them. It is at this point, also, that we begin to question whether schemes against surprise attack can be viewed as "first steps" toward more comprehensive disarmament in the traditional sense, or instead are incompatible with other forms of disarmament. Can measures to protect SAC be viewed as first steps toward its dismantlement? Can we initially take cooperative measures to perfect and safeguard each side's capacity to retaliate massively, in the interest of mutual deterrence, and do it as a step toward eliminating the threat of massive retaliation from a tense and troubled world?

Or should we instead recognize measures to safeguard against surprise attack as a compromise — an implicit acceptance of "mutual deterrence" as the best source of military stability we are likely to find — and a recognition that though we may not be able to replace the balance of terror with anything better, there may be much that we can do to make that balance stable rather than unstable.[3]

Once we have identified the surprise-attack problem as the possible vulnerability of either side's retaliatory force to a first strike by the other, it becomes necessary to evaluate military strength, defensive measures, and proposals for the inspection or limitation of armament, with precisely this type of strategic vulnerability in mind. We do not, for example, assess American and Soviet strategic forces by counting up the bombers, missiles, submarines, and aircraft carriers on both sides, as though we wanted to see who could put on the most impressive peace-time parade. "Who is ahead" in the arms race will usually be: *whoever strikes first*. And if we have to plan on the conservative assumption that the other side will strike first, 200 bombers safe against attack may be worth as much as 2000 that have only a 10 per cent chance of survival.

An assessment of defensive measures also comes out differently if we put primary reliance on deterrence. Chicago cannot be hidden, buried in a blast-proof cavern, or kept 10 miles off the ground; but concealment, dispersal, hard shelter, and airborne alert are meaningful defenses in preserving the deterrent force. An active air defense of Chicago that has only a 50-50 chance

---

[3] In case the reader feels that the argument presented here is correct in principle but uninteresting in fact because the continuous invulnerability of our retaliatory forces is assured beyond any worry, I should like to refer him to Albert Wohlstetter's cogent discussion in "The Delicate Balance of Terror," *Foreign Affairs*, 37:211–234 (January, 1959).

of saving the city from a multi-megaton bomb would be a discouraging prospect, and we have little promise that we could even do that well; but an active defense that could guarantee the survival of a large fraction of our strategic striking force might be more than enough to guarantee the Russians a prohibitive cost in retaliation. Similarly, a defense of Chicago that requires the enemy to triple the size of his attack may be a poor prospect; it may mean only that he invests in a larger initial attack. But a defense of our retaliatory force that requires the enemy to triple the size of his attack may substantially increase the enemy's difficulty of sneaking past our warning system, and appreciably change his likelihood of successfully precluding retaliation.

The same kind of calculation is pertinent to an evaluation of arms limitations. If we look only at the problem of a Russian attack on American cities, it may seem immaterial to the enemy whether he shoots his ICBM's from close up or from afar; accuracy may not make much difference with a multi-megaton bomb fired at metropolitan areas. But if he is trying to destroy a missile or bomber that has been sheltered deep underground with reinforced concrete, accuracy is no longer superfluous. An average aiming error of two or three miles may be nothing in shooting at a large metropolitan area; an attempt to knock out a hard-sheltered retaliatory weapon may require several missiles to get a direct enough hit. Thus *zonal limitations* on the placement of ICBM's might seem an ineffectual form of disarmament in the conventional sense; but in stabilizing deterrence — in reducing the vulnerability of each side's retaliatory forces to the other's forces — the separation of each side's missile sites from the other's, by reducing accuracy, might make a real difference. (For unsheltered planes or missiles, of course, the city-target analogy is unfortunately pertinent.)

On some questions, emphasis on the surprise-attack problem may lead to a downright reversal of the answer that one would get from more traditional "disarmament" considerations. Consider the case of a limitation on the number of missiles that might be allowed to both sides (if we ever reached the point in negotiations with Russia where an agreement limiting the number of missiles were pertinent and inspection seemed feasible). Suppose we had decided, from a consideration of population targets and enemy incentives, that we would need a minimum expectation of 100 missiles left over after his first counter-missile strike in order to carry out an adequately punitive retaliatory strike — that is, to deter him from striking in the first place. For illustration suppose his accuracies and reliabilities are such that one of his missiles has a 50-50 chance of knocking out one of ours. Then, if we have 200, he needs to knock out just over half; at 50 per cent reliability he needs to fire just over 200 to cut our residual supply to less than 100. If we had 400, he would need to knock out three-quarters of ours; at a 50 per cent discount rate for misses and failures he would need to fire more than twice 400, that is, more than 800. If we had 800, he would have to knock out seven-eighths of ours, and to do it with 50 per cent reliability he would need over three times that number, or more than 2400. And so on. The larger the initial number on the "defending" side, the larger the *multiple* required by the attacker in order to reduce the victim's residual supply to below some "safe" number.[4]

From this point of view, a limitation on the number of missiles would appear to be more stabilizing, *the larger the number permitted*. This would be so for two reasons. First, the larger the number on both sides, the greater is the absolute number of missiles expected to be left over for retaliation in the event that either side should strike first, and therefore the greater is the deterrence to an attempted first strike. Second, the larger the number of missiles on both sides, the greater must be the absolute and proportionate increase in missiles that either side would have to achieve in order to be capable of assuring, with any specified probability, that the other's left-over missiles would be less than some specified number after being attacked. Thus the difficulty of one side's cheating, by disguising and concealing extra missiles, or breaking the engagement and racing to achieve a dominant number, is more than proportionately enhanced by any increase in the starting figures on both sides. In fact, if the numbers to begin with are high enough to strain the budgetary

---

[4] This assumes that he fires his missiles all together or that, if he fires successive salvos, he has no means of reconnaissance that lets him know, on successive salvos, which particular missiles have already destroyed their targets.

capacities of the two enemies, and within these budgetary capacities the number of missiles is high, stability might be imposed by the economic limitation on what either side could do relative to what it would have to do to achieve mastery.

Here is a case, then, in which an "arms race" does not necessarily lead to a more and more unstable situation. For anything like equal numbers on both sides, the likelihood of successfully wiping out the other side's missiles becomes less and less as the missiles on both sides increase. And the *tolerance* of the system increases too. For small numbers on both sides, a ratio of 2 or 3 to 1 may provide dominance to the larger side, a chance of striking first and leaving the other side a small absolute number for striking back. But if the initial numbers on both sides are higher, it may take a ratio of 10 to 1 rather than 2 or 3 to 1 to have a good chance of striking with impunity. Neither side needs to panic if it falls behind a little bit, and neither has any great hope that it could draw far enough ahead to have the kind of dominance it would need.

This greatly simplified view of a "missile duel" is much too specialized to be a strong argument for arms races rather than disarmament. But it does demonstrate that, within the logic of stable deterrence, and of schemes for the prevention of surprise attack, the question of more vs. fewer weapons has to be analyzed on its merits in individual cases. It is *not* a foregone conclusion that disarmament, in the literal sense, leads to stability.

Our attitude toward missile submarines, and toward the problem of devising submarine-detection techniques, should be much affected by whether we are worried about enemy attack or enemy *surprise* attack. If the submarine proves to be for many years a fairly invulnerable site for anti-population missiles, we should perhaps view it not as an especially terrifying development but as a reassuring one. If in fact the best we can hope for is mutual deterrence and we only want the balance to be stable, then the polaris-type missile carried by a submarine of great mobility and endurance may be the kind of weapon system that we should like to see in adequate numbers on both sides. If it should prove to be both undetectable and highly reliable, it would have the advantage of not needing to strike first in order to strike at all, of not fearing that an aggressor might hope to knock out the very forces that were supposed to deter him. True, it might seem more reassuring if we had the power to destroy the enemy's missile subs while he did not have the power to destroy ours; but if the power already exists on both sides, and we cannot wish it away, then the most we can hope for is that this capacity to destroy each other be itself sufficiently indestructible that each side is in fact deterred. From that point of view, we perhaps should not even wish that we alone could have the "invulnerable" nuclear-weapon submarine; if in fact we have either no intention or no political capacity for a first strike, it would usually be helpful if the enemy were confidently assured of this. His own manifest invulnerability to our first strike could be to our advantage if it relieved him of a principal concern that might motivate him to try striking first. If *he* has to worry about the exposure of *his* strategic force to a surprise attack by *us, we* have to worry about it too.

These thoughts also affect our attitude toward the search for submarine detection. The Navy is urgently seeking a better system of defense against submarines, and there is no question but that we have to devote ourselves intently to the problem. Yet perhaps we ought simultaneously to *hope* that the problem is insoluble. If it were insoluble (in the relative sense in which a technical problem can ever be insoluble) and submarines were destined to be comparatively safe vehicles for a decade or so, stable deterrence might be technologically possible. If submarines prove to be vulnerable themselves, arms technology is less stable than we hope. We have to *try* to detect submarines, because we cannot afford to let the Russians find a technique that we do not know, and because we have to learn all we can about detection to make our submarines less detectable; but like a person who has entered into an agreement with a partner that he cannot trust, we may search like the devil for a loophole, knowing that our partner is searching just as hard, while hoping that no loophole is to be found.[5]

---

[5] This paper being about principles, not about submarines, I can perhaps be excused for pretending here that undetectability on short notice in the open sea is equivalent to invulnerability.

Once we have pressed the argument this far, we may as well carry it all the way. If our problem is to guarantee to an enemy that we have the ability to strike a punitive blow after being struck ourselves — and to assure him that we know that he knows it so that we are under no temptation to doubt the potency of our own deterrence and strike first — we should find virtue in technological discoveries that enhance the anti-population potency of our retaliatory weapons. If it is logical to take measures to guarantee that a larger proportion of our retaliatory forces could survive a first strike on them, the same logic should make us welcome an increase in the potency of those that do survive. As Bernard Brodie has said, "When we consider the special requirements of deterrence, with its emphasis on the punitive aspect of retaliation, we may find a need even for super-dirty bombs. Since the emphasis must be on making certain that the enemy will fear even the smallest number of bombs that might be sent in retaliation, one wants these bombs to be, and thus to appear before the event, as horrendous as possible." [6]

The novelty of this reasoning disappears as soon as we recognize that the "balance of terror," if it is stable, is simply a massive and modern version of an ancient institution: the exchange of hostages. In older times, one committed himself to a promise by delivering his hostages physically into the hands of his distrustful "partner"; today's military technology makes it possible to have the lives of a potential enemy's women and children within one's grasp while he keeps those women and children thousands of miles away. As long as each side has the manifest power to destroy a nation and its population in response to an attack by the other, the "balance of terror" amounts to a tacit understanding backed by a total exchange of all conceivable hostages. We may not, of course, want to exchange quite *that* many hostages in support of this particular understanding with this particular enemy. But in a lawless world that provides no recourse to damage suits for breach of this unwritten contract, hostages may be the only device by which mutually distrustful and antagonistic partners can strike a bargain. [7]

[6] Bernard Brodie, *Strategy in the Missile Age* (Princeton, 1959), p. 295.
[7] It should be emphasized that I am discussing only the problem of major

This line of reasoning is not simply an enormous rationalization for an arms race. It does indeed suggest that "disarmament" in the literal sense, aimed indiscriminately at weapons of all kinds — or even selectively aimed at the most horrifying weapons of mass destruction — could produce instability rather than stability, and might have to be *completely* successful in order not to be disastrous. Nevertheless, there is an important area of arms limitations that is not only compatible with the foregoing analysis but is suggested by it.

It suggests making a distinction between the kinds of weapons that are peculiarly suitable to the exploitation of a first strike and weapons that are peculiarly suitable to the retaliatory role. At one extreme is the "pure" strike-back type of weapon: the relatively inaccurate vehicle with a super-dirty bomb that can kill just about everything in the enemy's country *except* a well-protected or well-hidden retaliatory force, and that itself is so well-protected or well-hidden as to be invulnerable to any weapons that the other side might possess. Ideally, this weapon would suffer no disadvantage in waiting to strike second and gain no advantage in striking first. At the opposite extreme is a weapon that is itself so vulnerable that it could not survive to strike second, or a weapon so specialized for finding and destroying the enemy's retaliatory forces before they are launched that it would lose most of its usefulness if it were held until the other side has already started. These "strike-first" weapons not only give their possessor a powerful incentive to strike first, and an incentive to jump the gun in the event of ambiguous warning rather than to wait and make absolutely sure; they are a tacit declaration to the enemy that one expects to strike first. They consequently invite the enemy to strike a little before *that* and to act with haste in the event he thinks that we think it's time to act quickly.

Between the extremes of the "pure" strike-first weapon and the

---

surprise attack here. The implications of the "hostage" concept for, say, civil defense policy depends on its relation to other contingencies as well — e.g., limited war, mischief by a third party, less-than-massive retaliation, etc. One of these interrelations between surprise attack and other military contingencies is touched on in the final pages of this chapter.

"pure" strike-back weapon, there are the weapons that can strike first but do not need to, that can survive and serve the retaliatory purpose but that also might have an important effect on the other side's retaliatory forces if used first. Perhaps most weapons fall in this category if reasonable precautions are taken for their protection. So we cannot make a nice distinction between first-strike and second-strike weapons, extolling the one and disparaging the other in our approach to the surprise-attack problem. If we were to consider eliminating all weapons that had any possible effect against the other side's retaliatory forces, or that enjoyed any advantage in being used first, there might not be enough left with which to promise retaliation.[8] But surprise-attack negotiations might usefully concentrate on the opposite extreme.

The most obvious candidates would be exposed, vulnerable weapons. It might seem anomalous to insist to the Russians that they cover any nakedness of their strategic forces, or for them to suggest that we protect better some of our own. More likely would be suggestions to abandon weapons that were provocatively exposed to the other side. Note how different in spirit this would be from the "ban the bomb" orientation. Whatever the propaganda implications of such a topic, it at least has the merit of viewing deterrence as something to be enhanced, not dismantled.

Second, restrictions on the deployment of forces that affect their counter-force potency rather than their counter-population potency might be sought. They will not be sought, however, until there is candid recognition that surprise-attack schemes are to be deliberately aimed at protecting, not degrading, each side's strike-back capability. The discussion above of the effect of range on missile requirements, whatever its specific merits, suggests that this class of limitations is not an empty one.

Third, there may be some useful exploration of cooperative measures, or mutually accommodated modes of behavior, that reduce the danger of war by misapprehension. Even voluntary exchange of information might help, if we and the Russians can unilaterally pick modes of behavior that, when the truth is

---

[8] Furthermore, we are taking nothing but the surprise-attack problem into account here.

known, are reassuring. This is presumably the idea behind proposals for inspection of air traffic in the north polar area, and there may be some other types of activity in which there could be mutual benefit from some traffic rules. What is attractive about these measures — as about a candid discussion of the evils of strike-first weapon systems — is that they may make possible some understandings that do not have to be embodied in formal agreements, and may facilitate unilateral accommodations on both sides.

Fourth, there may be arrangements to cope with crises and emergencies that threaten to explode into an unintended war. A later section of this chapter discusses this point at some length.

Fifth, there may be measures that, by making surprise less likely, make a first strike less attractive. This point brings us back to the open-skies type of proposal.

Most public discussion of the surprise-attack problem during the last few years has related to measures that might reduce the likelihood of surprise, rather than measures to limit what weapons could do if surprise were achieved. The open-skies proposal was based on the idea that with sufficient observation of each other's military forces neither side could achieve surprise and, lacking the advantage of surprise, would be deterred.

The technical problem of devising a practical inspection scheme that could yield each side adequate warning of an attack by the other has become much more difficult since the first open-skies proposal was made. With hydrogen weapons reducing the number of aircraft that might be needed in a surprise attack, with missiles promising to reduce the total time available between the initial actions in readying a strike and the explosion of weapons on target, and with mobile systems like missile submarines to keep under surveillance, it looks as though pure inspection unaccompanied by any limits on the behavior of the things to be inspected would be enormously difficult or enormously ineffectual. The idea of examining photographs for strategic indications of force movements and concentrations is simply obsolete. The problem now would seem to be one of intensive surveillance of strategic forces by a vast organization that could transmit authentic messages reporting suspicious activity within at most a few hours, and eventually within a few minutes, in a way that is not intolerably susceptible of false alarms. There is no practical assurance that this could be done.

This does not mean that inspection schemes against surprise attack have no prospect of success. What it means is that a scheme providing for *nothing but inspection* may have very poor prospects. But if one cannot send observers out to follow all the aircraft, missiles, and submarines wherever they go, one can still consider calling the aircraft, missiles, and submarines to assemble where they are more easily watched. If restrictions on the deployment of forces are used to make the task of inspection more manageable, something may be accomplished. But though there may be promise in the idea of combining inspection and weapon limitations, there are also serious problems.

One is a possible incompatibility between the need for inspection and the need for concealment. When missiles become sufficiently accurate, it may become almost physically impossible to protect one's own retaliatory forces by the sheer provision of cement, or, if not impossible, exceedingly costly. Mobility and concealment may then have to be the source of security for the retaliatory forces; if the enemy can hit anything he can locate, and kill anything he can hit, he has to be made unable to locate it. To the extent that he can have our own retaliatory weapons under continuous surveillance he has continuous information on their location.

In other ways an inspection scheme on the scale required for protection against surprise attack might yield excessive information about the disposition of the other's forces and make them more vulnerable. It is widely known, for example, that there was a time when hurricane winds immobilized an extremely large portion of the B-36's that then comprised our principal retaliatory threat. The implications for surprise attack of such an event are evidently very different, depending on whether the enemy knows only in a general way that this kind of thing can happen to us, or instead has definite information when it occurs and knows exactly whether or not he has clear sailing for a few days. Imagine the state of tension that could occur if *either* side's strategic-force personnel began to suffer a severe epidemic that threatened to immobilize them temporarily before the eyes of the other side's inspection. Much better — if we and they are occasionally to land in a very unalert position for reasons that are impossible to prevent — that neither of us should be in a position to know too much about the other's occasional disabilities.

Finally, while there may be arrangements that have a high probability of providing warning of the enemy's preparation for an attack, the value of the system depends on what we can do if we do get warning. We can send off our own anticipatory strike, hoping to get in first; but this is an unattractive course if the warning is ambiguous. A false alarm then leads to war. And a true one precludes any last-minute deterrence.

At the other extreme we can just wait and "get ready." And if the things we can do to get ready appreciably reduce the likelihood that his attack will succeed — if they raise the likelihood that we can retaliate severely — we may want to make a quick demonstration to the enemy that we are ready, in the hope that our improved posture will deter his final decision.

The important question is what we do that constitutes getting ready. If the answer is simply, "Be more alert," why weren't we more alert in the first place? Most of the obvious things that one would do if he had warning of an attack are things that one probably would like to do perpetually in view of the ever-present possibility of an attack. And if our Strategic Air Command is continually doing its best to reduce the time it takes to get aircraft ready and off the ground in the face of warning, or to keep the doors tightly shut on sheltered aircraft, or to keep aircraft safe in the air in combat-ready condition, there may not be much more they can do on short notice.

Nevertheless, there are things that a nation can do in the face of imminent attack that it could not do continuously and indefinitely. One can evacuate or go underground, but not forever. One can get his retaliatory forces safely off the ground, where they are no longer targets for enemy bombs; but they cannot stay in the air forever. One can put men on twenty-four-hour duty, but not for many days in a row. One can ground all commercial aircraft to raise the reliability of the warning system, but the economic loss might be exorbitant if commercial and private flying were foresworn for all time in the interest of making enemy aircraft more recognizable. There are, in other words, things that one can do to "get ready" in the face of expected attack that one cannot be expected to do continuously.

But there is another question. How long can we keep it up? Suppose we cannot physically keep all aircraft in the air at all times, as is true, and that it may be too costly in all respects (accidents as well as fuel and crews) to keep as many as half of them in the sky on the average, but that a substantial increase in the number aloft can be affected on short notice if a serious warning is received. This might well mean that the enemy would not be deterred by our ordinary posture, but would be deterred by the posture we can adopt when we get warning. Does this mean that he quits as soon as he sees that we are ready? Or might he just wait until the gas is gone, the pilots are tired, and the planes have to come down again? And if so, must we not strike in anticipation?

This problem of "fatigue" is likely to plague any super-alert stance that one can take. The solution is in two parts. First, one must try to design a super-alert response that has good endurance and little fatigue, recognizing that this means compromising its peak effectiveness. Second, and most pertinent to the present subject, one may have to engage in a kind of crash disarmament negotiation with the enemy during the period that one has in fact taken measures to insure his own invulnerability of retaliation. If we can keep up a super-alert for a few days, we have a few days during which to attempt to demand or negotiate some degree of Russian "disarmament" that is both tolerable to them and sufficiently reassuring to us to permit us to return to "normal" rather than to proceed with total war. This might mean devising and instituting a much more ambitious scheme of anti-surprise-attack measures than had been politically feasible during the earlier period. It would mean negotiating not just under the ordinary pressure of knowing that sneak attack is a long-term danger, but doing it with clear notice that if measures to make successful first-strike impossible have not been devised, agreed upon, and taken by a quick deadline, war by mutual consent has become inevitable.

These reflections do not imply that extra warning would be either useless or embarrassing. What they indicate is that warning by itself may not be enough. Extra warning provides an *opportunity*, but the opportunity has to be exploited with skill. And preparations for what one would do in the contingency may have to be made well ahead of time. There is barely time to deliver an ultimatum to the Russians when we catch them preparing to attack. Deciding what ultimatum would both meet our needs and be tolerable to the Russians is not only intellectually difficult, it is technically difficult, depending on such things as procedures to verify compliance. We could probably deliver an effective ultimatum only if we had planned carefully ahead of time on what it might contain.

There are two quite distinct criteria for judging the efficacy of an inspection system, or for designing the system itself. One is how well the system gets at the truth in spite of efforts to conceal it; the other is how well it helps one to reveal the truth convincingly when it is in his interest to do so. The difference is like that between a scheme for discovering the guilty and a scheme for permitting the innocent to establish innocence. Roughly speaking, one system arrives at a presumption of innocence in a negative way, by an absence of positive evidence to the contrary; the other scheme relies on positive evidence, and is pertinent to the particular situations in which one's own interest is in letting the truth be known.

The difference between these two situations is pertinent to the distinction between a scheme to minimize the fear of *deliberate* surprise attack and a scheme to minimize the fear of inadvertent, or "accidental," or unintended war — the war that results from a false alarm, or from a mistaken evaluation of the other's response to a false alarm, or to a wrong interpretation of a mechanical accident, or to the catalytic mischief of a third party interested in promoting war, or to a situation in which the apprehension by each side that the other may be about to pre-empt explodes by feedback into a war by mutual panic. In the case of a planned, deliberate, surprise attack, the aggressor has every reason to disguise the truth. But in the case of "inadvertent war," both sides have a strong interest in conveying the truth if the truth can in fact be conveyed in a believable way in time to prevent the other side's mistaken decision.

MISAPPREHENSION OF ATTACK

Consider this question: how would we prove to the Soviet Union that we were not engaged in a surprise attack, when in fact we were not but they thought we might be? How might they prove to us that they were not initiating a surprise attack, if in fact they were not but they knew that we were afraid they might be.

Evidently it is not going to be enough just to tell the truth. There may indeed be some situations in which sheer verbal contact is enough to allay each side's suspicions. If the Russians — just to take a wild example — suffered an accidental nuclear explosion on one of their own bases, it might be helpful to both sides if they could simply reassure us quickly that they knew it was an accident, that they were not interpreting it as a harbinger of an attack by us, and so on. But, in most of the cases that one can imagine, it is insufficient simply to assert that one is not engaging in a strategic strike or that one is not in a menacing posture. There has to be some way of authenticating certain facts, the facts presumably involving the disposition of forces. We would have to prove not only that we were not *intending* to exploit our position, but that our actual position was one that *could not* be exploited to doublecross the enemy if he should take us at our word and restrain his own forces.

MISAPPREHENSION DURING LIMITED WAR

Especially in the course of a limited war one side or the other may take an action that might be misinterpreted as a strategic strike. Suppose, for example, that we used the kinds of aircraft that would alternatively be used in a strike against Russian bases, and flew them in directions that might be interpreted as aimed at the Soviet Union itself — as might be the case if they. were flying from North African bases or the Mediterranean fleet to countries near the southern border of the Soviet Union. Alternatively, suppose that the Soviet aircraft flew a limited war mission that could be interpreted, on the basis of the momentary evidence we might get, as a strike at all of our overseas bases and carriers, but that was actually a limited strike and not part of a general effort to destroy United States retaliatory power.

The question arises whether there are any means by which to reduce the likelihood of misinterpretation in this case, where misinterpretation might lead one side either to take off in anticipated retaliation, to pre-empt as quickly as it could, or to get into a super-alert status that had a high proclivity toward false alarm. One might wish to bend over backwards to demonstrate that complementary actions — actions involving other forces in other parts of the world, that would almost certainly take place if this were an all-out counter-force strike — were in fact not being taken.

RECIPROCAL MISAPPREHENSION

Consider another case that was described by Gromyko at a press conference.

After all, meteors and electronic interference are reflected on Soviet radar screens, too. If in such cases Soviet aircraft, loaded with atomic and hydrogen bombs, were to proceed in the direction of the United States and its bases in other states, the air fleets of both sides, having noticed each other somewhere over the Arctic region, under such circumstances would draw the natural conclusion that a real attack by the enemy was taking place, and mankind would find itself involved in the whirlpool of atomic war.

Assuming for the moment that a situation like that described might conceivably arise, how might the interacting misapprehensions of both sides be slowed down and reversed? If there were some way of reversing motion on both sides, in a properly phased and authenticated way, a kind of balanced withdrawal by mutual consent might be possible.

The bargaining environment is not a propitious one. At best there would be only hours in which to conduct the negotiations, and at worst no time at all. The requirements for a successful outcome can analytically be divided into two parts. First there has to be discovered some "solution" — some pattern of action that reverses the trend toward mutual attack, and that constitutes a dynamically stable withdrawal to a less menacing alert status, one that yields neither side a dangerous advantage in the process, and that is within the physical capabilities of the forces concerned. The second requirement is that compliance somehow be observable, verifiable, and provable. We cannot carry out our part of the bargain unless we have trustworthy means for monitoring the other side's compliance, and the same is true for them. Conceivably we would have an interest in cheating; but it is overwhelmingly more probable that we should wish in these circumstances for a cheat-proof monitoring system that we could submit to, so that if we did comply with our part of the bargain the other side would have no doubt about it. The problem is essentially one of contract enforcement. And the motivation in this case, for each side, is to convey the truth as best it can if in fact it complies with the plan.

This example not only makes clear the need for some *prior arrangement* for observation and verification, in view of the very short time available for bringing inspectors to the scene; it also demonstrates how important it is to have thought ahead of time about what kind of proposal to make, and to have designed one's own flight plan in a way that could take maximum advantage of any means we might have for deliberately giving the enemy true information in the event it becomes desperately necessary to do so.

This case also may illustrate the difference between the two criteria for reliability of an inspection system. It might be very difficult to design radar that would *always* catch the enemy — and by which he could always catch us — in an attempt at sneak attack; it is quite another question how to design radar so that if we both wished to invite voluntary surveillance we could submit in a convincing way. In one case we are, in effect, evading his radar surveillance as best we can. In the other we may deliberately "parade" in front of his radar, or submit to other means of long-distance recognition, as long as he does the same for us.

LONGER-TERM SURVEILLANCE

The difference between these crises and emergency situations and the longer-term problems of policing arms limitations is in the kind of evidence that is required and in the strength of the motivation to provide it. The more "leisurely" process of inspection is generally viewed as depending mainly on *negative evidence,* that is, the *absence of evidence.* One reduces the probability of missing such evidence by enlarging and intensifying the system; and one supposes that the evasion is made difficult by the need to keep activities hidden over a long period. But in a crisis one requires more certain evidence; one does not have time to get leads and follow them up; there is no time to try the system out and enlarge it or intensify it if it does not work. Consequently, a crisis agreement would have to rely on *positive evidence.* Instead of looking for evidence about what the other party is *not* doing, one demands evidence that shows what he *is* doing. And the reason why such evidence might be forthcoming in a crisis is that the motive to provide it — the greater urgency of reaching an understanding or an agreement that depends on it — may be enhanced in such an emergency.

OVERBUILDING THE SYSTEM

For the purpose of being at least somewhat prepared for crises and unforeseen situations, there is a good argument for instituting some flexible stand-by arrangements for communicating with potential enemies and inspecting each other. In particular there is a good argument for overbuilding an inspection system relative to such use as has been agreed on. Having standby capacity to enlarge or intensify the system, or to augment it with additional facilities and inspectors, may have a good deal to do with the usefulness of the system in time of crisis. To put the point differently, we should not judge the reliability and usefulness of a system solely in terms of the motivations of the participants during "normal" operations; we should recognize that occasions may arise when there is a powerful motive for crash negotiations on arms limitations, at least momentary limitations, with no time available for setting up observation and communication systems *ad hoc*.

To be specific: in the event there should be established an inspection system to monitor an agreement to suspend nuclear tests, we should consider carefully how both sides might take advantage of the inspectors and their facilities in the event of an acute military crisis. The mobility of the inspectors, their location, their communication facilities, their technical training and surveillance equipment, there trustworthiness, and their numbers, should be evaluated and designed not just with nuclear-test detection in mind, but with some view to their serving a desperately critical need for a means of inspection, verification, and communication, in a crisis that threatens both us and the Russians with inadvertent war.

From the foregoing considerations, it is not at all clear that the stability of the balance of terror — the lack of temptation to deliberate surprise attack, and the immunity of the situation to false alarm — will be greatly affected by the military arrangements that we try to work out with the Russians. As nature reveals her scientific and technological secrets over the coming years, we may find that each side (if it does what it ought to do and does it rapidly enough) can substantially assure the invulnerability of its own retaliatory forces irrespective of what the other side does, and assure it in a convincing way, so that a powerfully stable mutual deterrence results. Alternatively, nature may have planted mischievous secrets ahead of us, so that we and the Russians continually find new ways to destroy retaliatory forces at a faster rate than we find new ways to protect them. There is only a hope — no presumption — that even with great ingenuity and the best of diplomacy we and the Russians could find cooperative measures to arrest a trend toward instability. So we may get stability without cooperation, or we may not find it even with cooperation. Still, some kind of cooperation with the Russians, or mutual restraint, formal or informal, tacit or explicit, may prove to make a significant difference in the stability of the balance of terror; and the stakes of course are very high. So although we cannot be sure that a deliberate policy of collaborating to make each side's retaliatory forces invulnerable would make any difference, we have to consider that it might and to ask ourselves whether in fact we should want a perfectly stable balance of deterrence if we had the option before us. Would we really be interested in a far-reaching and effective anti-surprise-attack scheme if we knew of one, and if we thought the Russians would accept it?

Although it would be comforting to know that the Russians could not be tempted into a deliberate planned sneak attack, and comforting to know that they were so sure we wouldn't try it that they would never need to jump the gun in panic, it can nevertheless be argued that our ability to deter anything but a major assault on ourselves depends at least somewhat on the Russian belief that we might be goaded into deliberate attack. The Russians might not believe this if their retaliatory forces were substantially invulnerable to a first strike by ours. It can be argued that except under the most extreme provocation we would shrink from any retaliatory strike that had no significant chance of eliminating or softening the Russian return strike. According to this argument, a pair of *invulnerable* SAC's is a pair of *neutralized* SAC's; and while that might be the best kind in a completely bi-polar world, it is a luxury that we could not afford in the existing world — a world in which there is a large "third area" in which we wish to deter Russian aggression by a threat more credible than that of mutual suicide.

Can we threaten to retaliate, not just to resist locally, if the Russians unquestionably possess the military capacity to return us a blow of any size they please? Have the strategic forces any role when each is invulnerable to the other, except to neutralize each other and to guarantee, by their joint existence, their joint disuse?

There is a role. Strategic forces would still be capable of carrying out "retaliation" in the punitive sense. If the threat of knocking out Russian or Chinese cities was originally thought to be potent because of the sheer pain, economic loss, disorganization, and humiliation that would be involved, and not mainly because the military posture of the enemy in the immediate area of his aggression would be greatly affected, the main ingredient of the threat would still be present even if the other side's SAC were invulnerable.

The threat of *massive* retaliation, if "massive" is interpreted to mean unlimited retaliation, does indeed lose credibility with the loss of our hope that a skillfully conducted all-out strike might succeed in precluding counter-retaliation. But if we were ever to consider limited or graduated reprisals as a means of putting pressure on the Russians to desist from actions intolerable to us, or to consider extending a limited local war inside Russian borders in a way that maintained the pretence of local military action but was really intended to work through the sanction of civilian pain and the threat of more, this kind of retaliatory action, and the threat of it, might enjoy increased credibility with a reduction in the vulnerability of both sides' strategic forces. It does, paradoxically, for the same reason that *all kinds of limited war might become less inhibited as the possibility of all-out surprise attack became unavailable.* The risk involved in a bit of less-than-massive retaliation should be less than it is now because the fear of an *all-out* strike in return should be a good deal less. The fear that our limited retaliation would be mistaken for the first step in the initiation of all-out war should be less; the Russians would have to believe that we were literally prepared for suicide to mistake our limited retaliation for the initial step in mutual obliteration.

This is not to argue that limited retaliation, entailing the risk, if not the certainty, of limited counter-retaliation, cannot lead to total destruction, either slowly or by explosion into greater and greater retaliatory strikes, or would not be frightful to contemplate even if kept limited. The problem of limiting a war of reprisal may be no easier than that of limiting local war, and it may be harder. The argument here, however, does not depend on making an exchange of limited punitive blows appear safe and attractive *compared with limited local war,* but safe enough and attractive enough compared with all-out war to be a credible threat (and not a called bluff) *in any case where we may have to rely on the threat of retaliation.*

The strategic forces would thus be "neutralized" only in respect of potential attacks on each other; they would still possess a punitive role that provides some basis for a deterrent threat. While the threat of *all-out* punishment may lose credibility with the achievement of invulnerability by both sides' retaliatory forces, the threat of limited retaliation may well gain it. Whatever the net effect, we cannot deprecate a world of invulnerable SAC's simply by reference to the need for third-area deterrence; it has to be demonstrated that one particular deterrent threat (the massive one) is more potent than the other (limited) one.

Only an extreme optimist can think that we may ever have a clear choice of accepting or rejecting a scheme that would guarantee to make both sides' retaliatory forces totally and continuously invulnerable. But this question of what would happen to third-area deterrence, and the limited-retaliation possibility that it calls to mind, are pertinent to the question of what we might let ourselves hope for.

APPENDICES

NUCLEAR WEAPONS AND LIMITED WAR

With the development of small-size, small-yield nuclear weapons suitable for local use by ground troops with modest equipment, and with the development of nuclear depth charges and nuclear rockets for air-to-air combat, the technical characteristics of nuclear weapons have ceased to provide much basis, if any, for treating nuclear weapons as peculiarly different from other weapons in the conduct of limited war. It has, of course, been argued that there are political disadvantages in our using nuclear weapons in limited war, particularly in our using them first. Even those who consider a nuclear fireball as moral as napalm for burning a man to death must recognize as a political fact a worldwide revulsion against nuclear weapons.

This Appendix is about another basis for distinguishing between nuclear and other weapons. It involves our relations with the enemy in the process of limiting war. In the interest of limiting war or of understanding limited war, it may be necessary to recognize that a distinction can exist between nuclear and other weapons even though the distinction is not physical but is psychic, perceptual, legalistic, or symbolic. That small-yield nuclears delivered with "pinpoint" accuracy are just a form of artillery, and consequently do not prejudice the issue of limits in war, is an argument based exclusively on an analysis of weapons effects, not on an analysis of the limiting process — of where limits originate in limited war, what makes them stable or unstable, what gives them authority, and what circumstances and modes of behavior are conducive to the finding and mutual recognition of limits. The premise of the "just-another-weapon" argument is that, if there is no compelling weapon-effects basis for a distinction between nuclears and other weapons, there is no basis at all that is pertinent to the limiting process.

Is not the same point involved in discriminating among the users of weapons? There is no more difference between Russians and Chinese than there is between nuclear and other weapons; similarly for the difference between Chinese and North Koreans, or between Americans and Nationalist Chinese, British and Jordanians, Egyptians and Algerians. Yet nationality has been an important distinction in the process of limiting war or destroying its limits. Similarly, there is little difference between the terrain a hundred miles north of the Soviet-Iranian border and the terrain a hundred miles south, or what lies above the Yalu and below it, or the two sides of the Greek-Yugoslav border. Yet boundaries like these play an important role in the limiting process, quite aside from any physical difficulty in the crossing of rivers or the scaling of mountains that happen to coincide with them.

One could reply that these are "legal" distinctions and that legal distinctions are real ones while those between nuclear and other weapons are fictitious. But they are not really legal; they are "legalistic." There is no legal authority that forces the participants in limited war to recognize political boundaries or nationalities; the Russians are not legally obliged to treat a modest penetration of their border as a qualitative change in the war — as a dramatic act discontinuous with action up to their border. The Chinese were not legally obliged to retaliate (rather than just to resist) if we deliberately crossed the Yalu River; they did not lose any legal right to deny trespass by admitting occasional thoroughfare. We are not legally obliged to take cognizance of Russian pilots if they participate in a limited war, or Russian "volunteers" in a Near Eastern ground army fighting against our side. The inhibition on the penetration of a border, or on the introduction of a new nationality into the conflict, is like that on the introduction of a nuclear weapon; it is the risk of enemy response. And an important determinant of enemy response is his appreciation of what he has tacitly acquiesced in if he fails to respond, or makes only an incremental response, to our symbolically discontinuous act.

What makes the Soviet or Chinese border a pertinent or compelling place to draw a line in the event of war in that area is
principally that there is usually no other plausible line to draw.
For Western troops to cross the Russian border is to challenge —
not physically but symbolically — the territorial integrity of the
USSR, and to demonstrate or at least to imply an intention to
proceed. Unless one can find some "obvious" limit inside that
border, such that it would be clear to the Russians where we intended to stop in the event that we cross the border, and such
that it would be obvious to us that there was a limit to how far
the Russians would let us advance if we did cross it and that the
Russians knew that we knew it, there is just no other stopping
place that can be tacitly acknowledged by both sides. Under the
circumstances for the USSR to accept the penetration of that
border without a dramatic retaliation of some sort would be to
admit that Soviet territory is fair game for a gradually expanding
war. The political boundary is therefore *useful* as a stopping
place, not legally mandatory; it is useful to *both* sides in default
of any plainly recognizable alternative, since both sides have an
interest in finding some limit. The border has a *uniqueness* that
makes it a plausible limit. It is one of the few lines — perhaps
the only line, but certainly one of the few — that one could draw
in the region that could be tacitly recognized by both sides as the
"obvious" geographical limit that both sides might observe. It
has a compelling *power of suggestion,* a claim to attention, the
denial of which might seem — in default of any plainly recognizable alternative — to be a denial of any limitation.

But, if political-boundary and nationality considerations still
seem to be legal, and therefore real, consider some other distinctions that are significant in the limiting process. We provided
much equipment but no manpower to the war in Indochina; we
provided equipment, leadership, and advice to the Greek troops
during the guerrilla war, but no combat troops. We provide direct
naval support to the Nationalist Chinese in the Straits of Formosa. It has been thought that we might have given air support
to the French and Vietnamese in Indochina, without appearing
to the Chinese and Russians to be as "involved" as if we had put
ground forces in.

An economist can argue — with the same persuasiveness as those who argue that "pinpoint"-delivered small-yield weapons are just another form of artillery — that equipment and manpower are fungible resources in a military campaign, that air intervention is not "really" different from ground intervention, that military intellect is as important as leg muscle for troops that lack leadership and planning skill. The controversy about redefinition of service functions in the light of modern weapons, and about the usefulness of defining military-service functions in terms of the means of locomotion, suggests that an air-ground distinction or a naval-ground distinction rests on nothing but tradition. But the point of all this is that, in limiting war, tradition matters.

In fact, what we are dealing with in the analysis of limited war is tradition. We are dealing with precedent, convention, and the force of suggestion. We are dealing with the theory of unwritten law — with conventions whose sanction in the aggregate is the need for mutual forebearance to avoid mutual destruction, and whose sanction in each individual case is the risk that to breach a rule may collapse it and that to collapse it may lead to a jointly less favorable limit or to none at all, and may further weaken the yet unbroken rules by providing evidence that their "authority" cannot be taken for granted.

What makes atomic weapons different is a powerful tradition that they *are* different. The reason — in answer to the usual rhetorical question — why we do not ban bows and arrows on the grounds that they too, like nuclear weapons, kill and maim people, is that there is a tradition for the use of bows and arrows, a jointly recognized expectation that they will be used if it is expedient to use them. There is no such tradition for the use of atomic weapons. There is instead a tradition for their nonuse — a jointly recognized expectation that they may not be used in spite of declarations of readiness to use them, even in spite of tactical advantages in their use.

Traditions or conventions are not simply an analogy for limits in war, or a curious aspect of them; tradition or precedent or convention is the essence of the limits. The fundamental characteristic of any limit in a limited war is the psychic, intellectual, or social characteristic of being mutually recognized by both sides as having some kind of authority, the authority deriving mainly from the sheer perception of mutual acknowledgement, of a "tacit bargain." And a particular limit gains in authority from the lack of confidence that either side may have in what alternative limits may be found if the limit is not adhered to. The rationale behind the limit is legalistic and casuistic, not legal, moral, or physical. The limits may correspond to legal and physical differences or to moral distinctions; indeed, they usually have to correspond to something that gives them a unique and qualitative character and that provides some focus for expectations to converge on. But the authority is in the expectations themselves, and not in the thing that expectations have attached themselves to.

Whether limits on the use of atomic weapons, other than the particular limit of no use at all, can be defined in a plausible way is made more dubious, not less so, by the increasingly versatile character of atomic weapons. It is now widely recognized that there is a rather continuous gradation in the possible sizes of atomic-weapon effects, a rather continuous variation in the forms in which they can be used, in the means of conveyance, in the targets they can be used on, and so forth. There seems consequently to be no "natural" break between certain limited uses and others. If we ask, then, where we might draw a line if we wished to limit somehow the size of the weapons, the means of conveyance, the situations in which or the targets on which they can be used, the answer is that we are — in a purely technical sense — free to draw a line anywhere we please. There is no cogent reason for drawing it at any one particular gradation rather than another. But that is precisely why it is hard to find a rationale for any particular line. There is no degree of use, or size of weapon, or number of miles, that is so much more plausible than other degrees, sizes, or distances that it provides a focal point for both sides' expectations. Legalistic limits have to be qualitative and discrete, rather than quantitative and continuous. This is not just a matter of making violations easy to recognize, or of making adherence easy to enforce on one's own commanders; it concerns the need of any stable limit to have an evident symbolic character, such that to breach it is an overt and dramatic act that exposes both sides to the danger that alternative limits will not easily be found.

The need for qualitatively distinguishable limits that enjoy some kind of uniqueness is especially enhanced by the fact that limits are generally found by a process of tacit maneuver and negotiation. They are jockeyed for, rather than negotiated explicitly. But if the two sides must strike a "bargain" without explicit communication, the particular limit has to have some quality that distinguishes it from the continuum of possible alternatives; otherwise there is little basis for the confidence of each side that the other acknowledges the same limit. Even a parallel of latitude, or an international date line, or the north pole, may have this quality when no other natural, plausible, "obvious" point or line is available for expectations to converge on.

A test of this point with respect to atomic weapons might be to pose the following problem.[1] Let any of us try to cooperate for a prize: we are to sit down right now, separately and without any prior arrangements, and write out a proposed limitation on the use of nuclear weapons, in as little or as great detail as we please, allowing ourselves limitations of any description that appeals to us — size of weapons, use of weapons, who gets to use them, what rate or frequency of use, clean versus dirty, offensive versus defensive use, tactical versus strategic, on or not on cities, with or without warning — to see whether we can all write the *same* specification of limit. If we are in perfect agreement on the limits we specify, we get a prize; if our limits are different, we get no prize. We are doing this only for the sake of the prize, to see whether we can in fact agree tacitly on a statement of limits, and to see — for those of us who do manage to coordinate our proposals tacitly — what kinds of limits appear to be susceptible of tacit joint recognition. We are permitted the extremes of no limits at all on the one hand, or no atomic weapons at all on the other, and any gradation or variation defined in any way we please.

My argument is that there are particular limits — simple, dis-

---

[1] Compare Chapter 3, especially pp. 58–67.

crete, qualitative, "obvious" limits — that are conducive to a concerted choice; those who specify other kinds of limits, I predict, can find few partners or none at all whose limits coincide with theirs. (Since our object is to agree, we are to take no consolation in the other virtues of our proposed limits; in this exercise the main consideration in chosing any particular limits is the likelihood that if we chose those limits in an effort to coincide exactly with the limits of the others, knowing that they were trying to coordinate theirs with ours, we would succeed.)

I do not allege that this exercise proves what kinds of limits are capable of possessing stability and authority. It does demonstrate that certain characteristics of limits, particularly their simplicity, uniqueness, discreteness, susceptibility of qualitative definition, and so forth, can be given an objective meaning, one that is at least pertinent to the process of tacit negotiation. It suggests that certain kinds of limits are capable of being jointly expected by both sides, of focusing expectations and being recognized as qualitatively distinct from the continuum of possible alternatives.

The first conclusion to be drawn from this line of argument is that there is a distinction between nuclear and nonnuclear weapons, a distinction relevant to the process of limiting war. It is a distinction that to some extent we can strengthen or weaken, clarify or blur. We can strengthen the tradition, and enhance the symbolic significance of this distinction, by talking and acting in a way that is dramatically consistent with it; we can erode the distinction — but not readily destroy it — by acting as though we do not believe in it, by emphasizing the "just-another-weapon" argument and by making it evident that we in fact have little compunction about using nuclears. Which policy we should follow depends on whether we consider the distinction between nuclear and other weapons to be an asset that we share with the USSR, a useful distinction, a tradition that helps to minimize violence — or instead a nuisance, a propaganda liability, a diplomatic obstruction, and an inhibition to our decisive action and delegation of authority. Those who believe that atomic weapons ought to be used at the earliest convenience, or whenever military expedience demands, should nevertheless recognize the distinction that exists so that we can take action to erode the distinction during the interim.

This is not just a matter of what the Asian neutrals or our European allies feel about the distinction. It concerns a relation between us and the Russians — an understanding that may exist between us whether we like it or not. It has to do with whether the Russians think we share with them a tacit expectation that there is a limit against the use of nuclear weapons. In the interest of limiting war, we should want the Russians or the Chinese not to believe that our initial use of atomic weapons in a local war were a challenge to the whole idea of limitations, a declaration that we would not be bound by any kinds of limits. We should want them to interpret our use of nuclear weapons as consistent with the concept of limited war and consistent with our willingness to collaborate tacitly in the discovery and recognition of limits; we should want our use of atomic weapons not to be charged with excessive symbolic content. So, if I am right that a distinction does exist in the sense pertinent to the limiting of war, and if nevertheless we want maximum freedom to use atomic weapons, we ought in the interest of limiting war to destroy or to erode the distinction as best we can. (For example, a deliberate program for early and extensive use of "nuclear dynamite" in earth-moving projects, especially in underdeveloped countries, might help to erode the distinction; the same might be true of a program for training friendly troops in underdeveloped countries in how to survive nuclear weapons explosions, using some actual weapons for the purpose in their own country.) If on the contrary we wish to enhance the tacit understanding we have with our enemies that nuclears are a class apart and subject to certain reservations, agreement on nuclear test suspension (or even just extensive discussion of such an agreement) will probably contribute to the purpose.[2]

A second conclusion is that the principal inhibition on the use of atomic weapons in limited war may disappear with their first

---

[2] On the symbolic significance of a test agreement, see Henry A. Kissinger, "Nuclear Testing and the Problem of Peace," *Foreign Affairs*, 37:1–18 (Oct. 1958), especially pp. 12–13.

use. It is difficult to imagine that the tacit agreement that nuclear weapons are different would be as powerfully present on the occasion of the *next* limited war after they had already been used in one. We can probably not, therefore, ignore the distinction and use nuclears in a particular war where their use might be of advantage to us and *subsequently* rely on the distinction in the hope that we and the enemy might both abstain. One potential limitation of war will be substantially discredited for all time if we shatter the tradition and create a contrary precedent. (There may also be some limits or sanctuary concepts that we take for granted that should be reexamined to see whether they were originally by-products of the assumed nuclear ban and might disappear with it. We may want to look again at the role of naval vessels, for example, partly to anticipate enemy treatment of them, partly to avoid misinterpreting enemy intentions if he treats them differently after nuclears are brought into play.)

A third conclusion is that on the occasion of their first use we should perhaps be at least as concerned with the patterns and precedents that we establish, and with the "nuclear role" that we adopt, as with the original objectives of the limited war. For example, if nuclear weapons were used in defense of Quemoy, we probably ought to be much less concerned about the outcome on Quemoy than about the character of the nuclear exchange, the precedents that it establishes, the role we manage to assume for ourselves, and the role the enemy assumes in the process. We shall be not only using them *ad hoc* for the little war in question, but importantly shaping the limited nuclear wars to come. (When a boy pulls a switch-blade knife on his teacher, the teacher is likely to feel, whatever the point at issue originally was, that the overriding policy question now is his behavior in the face of a switch-blade challenge.)

Fourth, we should recognize that — at least on the first occasion when nuclear weapons are used in limited war — the enemy too will really be engaged in at least two different kinds of limited-war activity at the same time. One will be the limited struggle over the original objectives; the second will be the tacit negotiation or gamesmanship over the role of nuclear weapons themselves. To illustrate, we might in connection with Quemoy decide to use nuclear weapons; ordinarily it would be supposed that we should do this only if it were quite necessary to the defense of Quemoy, and that we should use them in a manner that achieves our Quemoy objectives. But, in considering whether the Chinese or Russians would use them in return, we should perhaps not worry mainly about what they think their use of nuclear weapons would do for the invasion of Quemoy. Much more important to them, it seems, would be the nature of their "response" to our nuclear initiative. They would be interested in not assuming a submissive role, but in demanding a kind of "parity" if not dominance in their own nuclear role. And, unless we are ready for some kind of decisive showdown in which we either win all or lose all, we must be as willing to "negotiate" (by our actions) for limited objectives in terms of nuclear dominance, traditions and precedents of nuclear use, and the "rules" we jointly create for future wars, as for any other types of objectives in limited war.

FOR THE ABANDONMENT OF
SYMMETRY IN GAME THEORY

The first part of this appendix argues that the pure "moveless" bargaining game analyzed by Nash, Harsanyi, Luce and Raiffa, and others,[1] may not exist or, if it does, is of a different character from what has been generally supposed; the point of departure for this argument is the operational meaning of *agreement*, a concept that is almost invariably left undefined. The second part of the paper argues that symmetry in the solution of bargaining games cannot be supported on the notion of "rational expectations"; the point of departure for this argument is the operational identification of irrational expectations.

A nontacit ("cooperative") nonzero-sum game — a bargaining game — is not *defined* by its payoff matrix; the operations by which choices are made must still be specified. Commonly these operations are sketched in by reference to the notion of "binding agreements" and the notion of free communication in the process of reaching agreement. Thus to say that two players may divide $100 as soon as they can agree on how to divide it, and that they may discuss the matter fully with each other, is generally considered sufficient to define a game.[2]

[1] John F. Nash, "The Bargaining Problem," *Econometrica*, 18:155–162 (April 1950), and "Two-Person Cooperative Games," *Econometrica*, 21:128–140 (January 1953); John Harsanyi, "Approaches to the Bargaining Problem Before and After the Theory of Games: a Critical Discussion of Zeuthen's, Hicks', and Nash's Theories," *Econometrica*, 24:144–157 (April 1956); R. Duncan Luce and Howard Raiffa, *Games and Decisions* (New York, 1957), pp. 114ff.

[2] Luce and Raiffa, in effect, *define* cooperative two-person games by reference to a payoff matrix and the following three stipulations. (1) All preplay messages formulated by one player are transmitted without distortion to the

A game of this sort is symmetrical in its move structure, even though it may be asymmetrical in the configuration of payoffs. The two players have identical privileges of communication, of refusing offers, and of reaching agreement. If instead of dividing $100 the players are to agree on values X and Y contained within a boundary, the payoff function may not be symmetrical but the move structure is. Harsanyi, to emphasize this, has even added explicitly the postulate of symmetrical moves: "The bargaining parties follow identical (symmetric) rules of behaviour (whether because they follow the same principles of rational behaviour or because they are subject to the same psychological laws)." [3]

What I want to do is to look at this notion of "agreement" on the assumption of *perfect symmetry in the move structure of the game,* paying close attention to the "legal details" of the bargaining process. We must also look at the meaning of "nonagreement." Since any well-defined game must have some rule for its own termination, let us look at the rules for termination first.[4]

If we are to avoid adding a whole new dimension to our payoff matrix, in the form of discount rates, we must suppose that the game is terminated soon enough so that nothing like the interest rate enters the picture. We do not want to have to consider the *time* at which agreement is reached, in addition to the agreement itself. This is more than a matter of convenience; the game ceases to be "moveless," except in very special cases, unless we make this stipulation. For, if the players' time preferences take any shape except that of a continuously uniform discount rate, the game itself changes with the passage of time and a player can, in effect, change the game itself by failing to reach agreement. The notion of a continuously uniform discount rate is probably far too special to treat as a *necessary* condition, and anyway has not been made

---

other player. (2) All agreements are binding, and they are enforceable by the rules of the game. (3) A player's evaluation of the outcomes of the game are not disturbed by these preplay negotiations. *Games and Decisions,* p. 114.

[3] John Harsanyi, "Approaches to the Bargaining Problem Before and After the Theory of Games . . . ," *Econometrica,* 24:149 (April 1956).

[4] The model discussed here is quite abstract, artificial, and unrealistic; but it does have the advantage of helping to test whether *even in an artificially abstract model* it is fruitful to postulate perfect symmetry in the move structure and to treat asymmetry as a special case, symmetry as the more general case.

an explicit postulate in the models under examination; so we must assume that the game is somehow gotten over with.

Perhaps the simplest way to terminate the game is to have a bell ring at a time specified in advance. There are other ways, such as having the referee roll dice every few minutes, calling off the game whenever he rolls boxcars. (We might have the game terminate after a specified number of offers have been refused, but this would change the character of the game by making certain kinds of communication "real moves" that leave the game different from what it was before, and perforce lead us into such tactics as the exhaustion of offers.)

For simplicity, suppose that the game will be terminated at a time specified in advance to the players, and for convenience let us call the final moment "midnight." If agreement exists when the midnight bell rings, the players divide the gains in the way they have agreed; if no agreement exists, the players receive nothing.

Next, what do we mean by "agreement"? For simplicity, suppose that each player keeps (or may keep) his current "official" offer recorded in some manner that will be visible to the referee when the bell rings. Perhaps he keeps it written on a blackboard that the other player can see; perhaps he keeps it in a sealed envelope that is surrendered to the referee when the bell rings; perhaps he keeps it punched into a private keyboard that records his current offer in the referee's room. When the bell rings, the blackboard is photographed, the envelope surrendered, or the keyboard locked, so that the referee needs only to inspect the two "current" offers as they exist at midnight to see whether they are compatible or not. If they are compatible, the gains are divided in accordance with the "agreement"; if the two players have jointly claimed more than is available, "disagreement" exists and the players get nothing. (Defer, for a moment, ruling on what happens if the two players together have claimed less than the total available, whether they get as much as they have claimed or get nothing for lack of proper agreement. And, in what follows, it will not matter whether an exhaustive agreement reached before midnight — that is, compatibility of the current offers occurring before midnight — terminates the game.)

There are other ways of defining "agreement" in terms of the operations by which it is reached or recorded; but if we adhere to the notion of a *perfectly symmetrical move structure* they will generally, I think, have the property that I am trying to single out for attention. That property is this. There must be some minimum length of time that it takes a player to make, or to change, his current offer. (For simplicity again, let us suppose that the same operation either makes an offer or changes it, so that we may always assume that a "current offer" exists.) There must then be some critical moment in time, a finite period before the midnight bell rings, that is the last moment at which a player can begin the operations that record his final offer. That is, there is some last moment before the bell rings, beyond which it is too late to change one's existing offer. Under the rules of the game and the rationality postulate both players know this. And by the rule of symmetry this moment must be the same for both players.

From this follows the significant feature. The last offer that it is mechanically and legally possible for a player to make is one that he necessarily makes without knowing what the other player's final offer is going to be; and the last offer that a player can make is one that the other player cannot possibly respond to in the course of the game. Prior to that penultimate moment, no offer has any finality; and at that last moment players either change or do not change their current offers, and whatever they do is done in complete ignorance of what each other is doing, and is final.[5]

This must be true. If either could get a glimpse of the other's final offer in time to do anything about it, or if either could give the other a glimpse of his own final offer in time for the other to respond, it is not — and is known to be not — a final offer.[6]

But now we have reached an important conclusion about the

[5] Incidentally, the argument is unaffected by supposing that a player can change his offer "instantaneously" as long as we keep the symmetrical rule that both can do it "equally instantaneously" as the final bell rings.

[6] There is a mechanical assumption here that in the process of making a new offer one can stop and start over. The case is slightly more complicated if an offer started one and one-half minutes before midnight is necessarily the last offer because the process cannot be started again until a minute has passed and by then the critical point has been passed. This case will be looked at again below.

perfectly move-symmetrical bargaining game. It is that it necessarily gives way, at some definite penultimate moment, to a *tacit* (noncooperative) bargaining game. And each player knows this.

The most informative way to characterize the game, then, is not that the players must reach overt agreement by the time the final bell rings or forego the rewards altogether. It is that they must reach overt agreement by a particular (and well-identified) penultimate moment — when the "warning bell" rings — *or else play the tacit variant of the same game.*

Each player must be assumed to know this and may, if he wishes, by simply avoiding overt agreement, elect to play the tacit game instead. So, if we assume (for the moment) that the tacit game has a clearly recognized solution, and that the solution is *efficient*, each player has a pure minimax behavior strategy during the earlier stage. Either can enforce this tacit solution by abstaining from agreement until the warning-bell rings; neither can achieve anything better from a rational opponent by verbal bargaining.

From this it follows that the solution of the cooperative game must be identical with that of the corresponding tacit game (if the latter has a predictable and efficient solution). It must be, because the tacit game comes as an inevitable, mechanical sequel to the cooperative game.

At this point it looks as though the cooperative feature of the game is irrelevant. The players really need not show up until 11:59; in fact they do not need to show up at all. The preplay communication and ability to reach binding agreements, which were intended to characterize the game, prove to be irrelevant; the cooperative game as a distinct game from the tacit game does not exist.[7]

But this conclusion is unwarranted. First, a tacit game may not

---

[7] In his 1953 article, "Two-person Cooperative Games," Nash presents a model that is explicitly tacit in its final stage. The model's relation to the cooperative game was heuristic: it was to help to discover what might constitute "rational expectations" (and hence the indicated rational outcome) in the corresponding cooperative game. The argument of the present paper is that the relation is likely to be mechanical rather than intellectual if a symmetrical move structure is strictly adhered to, and that with strict symmetry it is difficult, perhaps impossible, to define the corresponding nontacit game that was the ultimate subject of study.

have a confidently predicted efficient solution.[8] More than that, certain details of the cooperative game that might have seemed to be innocuous from the point of view of explicit negotiation may affect the character of the tacit game; similarly, preplay communication that has no binding effect on the players themselves may also affect the character of the tacit game. For an example, consider the following variant of the cooperative game.

Instead of saying that the players may divide a set of rewards if they can reach agreement on an exhaustive division, let us say that the players may divide a set of rewards *to the extent* that they have reached agreement on a division; they may divide such portion of the available rewards as they have already reached agreement on by the time the bell rings. If, for example, there are one hundred individual objects and the players have reached agreement on how to divide eighty of them when the bell rings, the twenty items in dispute revert to the house while the eighty on which agreement was reached will be divided in accordance with the agreement.[9]

[8] It should be emphasized that bargaining-game solutions that (like the Nash and Harsanyi solutions) depend on a clearly recognized zero point — that is, on an unambiguous outcome that reigns in the absence of overt agreement — cannot necessarily be applied to a cooperative game that is based on a matrix of choices. A matrix (unless perhaps all payoffs are zero except in the diagonal) does not have a zero point defined by the rules. There is consequently no "normal form" consisting of a convex region and associated zero point unless there is available a fully adequate theory that "solves" the tacit game (and does so in a manner that the players can take for granted). One may, following Luce and Raiffa (for example, page 137) take the players' "security levels" (maximin values) as the zero point; but this is either arbitrary or based on the hypothesis that, left to themselves, the players could succeed in doing no better than this in the tacit game. The latter hypothesis, especially where there are pure-strategy efficient points (as in Braithwaite's game, and as in the Luce-Raiffa matrix discussed in note 18 below), is a weak hypothesis that can be empirically refuted; it assumes that rational players are incapable of correlating strategies without communicating, while in fact this is something they often can do even in the face of conflicting preferences. (This point is taken up again in note 18.) The potential ambiguity of the zero point is the issue between Harvey Wagner and John Harsanyi in the former's, "Rejoinder on the Bargaining Problem," *Southern Economic Journal*, 24:480–482 (April 1958).

[9] In the case of a single divisible object like money, the corresponding rule might be that they divide the money in accordance with their offers after the house has removed the "overlap." Each player obtains as much as the other implicitly accords him; if one is demanding 65 percent of the money at the

Now, in the explicit-bargaining (cooperative) case, if we had already concluded there was an efficient solution to this game — that is, that the players would in fact reach an exhaustive agreement — we should probably have considered this reformulation of the problem inconsequential. The reformulation says, in effect, only that bargaining should take the form of each player's writing down the totality of his claim and that concessions shall take the form of each player's deleting items from his list of claims, with full agreement being reached when no more items are in conflict on the lists of claims. But, when we look at the tacit case, the game is drastically altered by this reformulation. The tacit game now has a perverse incentive structure. There is no rational reason for either player to demand less than the whole of the available reward; each knows this and knows that the other knows it. There is no incentive to reduce one's claim because any residual dispute costs the player no more than he would lose if he reduced his claim to eliminate the dispute. The single equilibrium point yields zero for both players. Thus the variant game, which seemed to differ inconsequentially, is drastically different from the original game; but it does not appear so until we have identified the terminal tacit game as a dominating influence.[10]

To take another example, suppose there are 100 individual objects to be divided and that, although they are fungible as far as value is concerned, the agreement must specify precisely which *individual items* go to which individual players. If the rules require that full and exhaustive agreement be reached, then in the tacit game the players are dependent on their ability not only to divide the total value of the objects in coordinated fashion but to

---

end of the game, and the other 55 percent, the second has been accorded 35 percent and the first 45 percent; these amounts are outside the range of dispute and constitute the "agreement."

[10] It might seem that we can draw a by-product from the analysis here, namely, the observation that in order to set up a "truly" cooperative (non-tacit) game, the legal definition of agreement must be such as to make the ultimate tacit game perverse, so that the players must reach binding agreement before the warning bell or suffer complete loss. But there is still a problem. The players themselves must now define "agreement" for purposes of their own agreement prior to the final bell. If it is like our earlier definition, all they accomplish is to make the perverse cooperative game into a benign one, one minute shorter, which is equivalent to a tacit game two minutes shorter than the original.

sort out the 100 individual objects into two piles in identical fashion. If, then, one of the players has demanded *specific* items worth 80 percent of the total and the other player has refused, the former has an advantage in the tacit game. The only extant proposal for dividing the 100 objects is the one player's specification of 80 that would satisfy him; the chances of their concerting identically on any other division of the 100 objects, equal or unequal between them, may be so small that they are forced for the sake of agreement into accepting the only extant proposal in spite of its bias. Thus preplay communication has tactical significance in that it can affect the means of coordination once the tacit stage of the game has been reached.

If now, in considering the tactical implications of this last point, we insist on a rule of symmetrical behavior, we must conclude that if either player opened his mouth to drown out what the other was about to say, he would always find the other player also with his mouth open, both knowing that if either spoke the other would be found to be speaking, neither able to hear the other, and so on. In other words, the assumption of complete symmetry of behavior as a recognized foregone conclusion seems to preclude the very kind of action that might have seemed to enrich the game at the stage of preplay communication.

But by now we have certainly pressed the perfect move-symmetrical game as far as is worthwhile.[11] We could go on to ana-

---

[11] One detail may be worth pursuing, in line with an earlier footnote. Suppose that it takes one minute to make or change an offer and (in contrast to the earlier version) that the process of recording a new offer, once started, cannot be stopped before it is completed. Under this procedure, any offer initiated during the next to last minute of the game is one's final offer. If this final offer *cannot* be communicated to the other player before the expiration of the minute, the game is essentially the same as before; "simultaneous" now means within a minute of each other for practical purposes, and again neither can see the other's final offer as he initiates his own, no matter what time during the final minute the offers are initiated. But suppose one punches his offer into a visible board which remains locked for one minute while the offer is recorded, so that the other player can see one's offer in a few seconds although one cannot initiate a change until the minute's delay is up. (And suppose that neither can make himself visibly incapable of seeing the other's offer once it is so recorded.) In this case, if the two offers during that final minute are not simultaneous, the player who moves second makes his final offer in full knowledge of the other's; and since his only chance of winning anything is to accept it, he must accept whatever the other has offered. Thus "second move" loses

lyze this game in more detail, considering such things as alternative ways of terminating the game or of defining "agreement," and so forth. It seems more worthwhile, however, to raise at this point the question of whether the perfectly "moveless" or "move-symmetrical" game is a profitable one to study. Is the nondiscriminatory, move-symmetrical game a "general" game, one that gets away from "special cases"? Or is it a special, limiting case in which the most interesting aspects of the cooperative game have vanished?

It should be emphasized that the fruitful alternative to symmetry is not the assumption of asymmetry, but just *nonsymmetry*, admitting both symmetry and asymmetry as possibilities without being committed to either as a foregone conclusion.

An illustration may help. Suppose we were to analyze the game in which there is $100 at the end of the road for the player who can get there first. This game of skill is not hard to analyze: the money goes to the fastest, barring accidents and random elements. We can predict rational behavior (running) and the outcome (money to the fastest). Ties will occasionally occur; but they will occur at the end of a race and will not be taken for granted at the outset. We need an auxiliary rule to cover ties, but it need not dominate either the game or the analysis.

Consider the same game played in a population in which everybody can run exactly as fast as anybody else, and everybody knows it. Now what happens? Every race ends in a tie, so the

---

if the first mover knows that the other is waiting. We now have a game that can be characterized as follows: the players dally around for 23 hours 58 minutes and then play a game lasting one minute, this game allowing each player one and only one offer which he can make at any time during the minute. This game offers, in effect, three strategies to a player, namely, (1) assume the other will wait, and demand 99 per cent; (2) assume both will make simultaneous offers, and demand whatever is indicated by the tacit game; (3) wait. If both wait, the game is still to be played. If there is a finite number of potential waits, we have strategies of wait-once-then-demand-99-per-cent, wait-once-demand-tacit-solution; wait-twice-demand-99-per-cent, wait-twice-demand-tacit-solution; and so on. This game (the "tacit supergame" consisting of all strategies for playing the one-minute game) is then *the* game; and it has, if we wish to accept it, its own "solution in the strict sense" which consists of all strategies (all lengths of waits) that end in demands that correspond to the solution of the tacit game. (For the definition of a solution in the strict sense in a tacit two-person game, see Appendix C.)

auxiliary rule is all that matters. But since a tie is a foregone conclusion, why would they bother to run?

The perfectly move–symmetrical cooperative game seems a little like that foot race. Bargaining in the one case is as unavailing as leg-work in the other; every player knows in advance that all moves and tactics are foredoomed to neutralization by the symmetrical potentialities available to his opponent. The interesting elements that we might inject in the bargaining game are meaningless if perfect symmetry, and its acceptance as inevitable by both players, are imposed on the game by its definition.

What should we add to the game to enrich it if the assumption of symmetry is dropped? There are many "moves" that are often available, but not necessarily equally available to both players, in actual game situation. "Moves" would include commitments, threats, promises; tampering with the communication system; invocation of penalties on promises, commitments, and threats; conveyance of true information, self-identification; and the injection of contextual detail that may constrain expectations, particularly when communication is incomplete. Such "moves" were discussed in detail in Chapters 2–5.

To illustrate, suppose in the earlier cooperative game there is a turnstile that permits a player to leave but not to return; his current offer as he goes through the turnstile remains on the books until the bell rings. Now we have a means by which a player can make a "final" offer, a "commitment"; whoever can record an offer favorable to himself and known to the other, and leave the room, has the winning tactic. Of course it may win for either of them; but this may mean that we end up with something like a foot race, and the one closest to the turnstile wins. By analyzing the tactic, and its institutional or physical arrangements, we may determine who can make first use of it.

We have not, it should be noted, converted the game of strategy into a game of skill by letting them race for the turnstile. It remains true that one wins when he gets to the turnstile first only through the other's cooperation, only by constraining the other player's choice of strategy. He does not win legally or physically by going through the turnstile; he wins *strategically*. He makes the other player choose in his favor. It is a tactic in a game of *strategy*, even though the *use* of it may depend on skill or locational advantage.

We can even put a certain kind of symmetry into the game now, without destroying it; we can flip a coin to see who is nearest the turnstile when the game begins, or let the players be similarly located and similar of speed but with random elements to determine who gets to the turnstile first. Though the *game* is now *nondiscriminatory*, the *outcome* would still be *asymmetrical* because each player has an incentive to run to the turnstile, leaving behind a standing offer in his own favor.[12]

We can include some risk of "tie," especially if there are two turnstiles and the players might go through them simultaneously. This constitutes "symmetry" as an interesting possibility, but not as a foregone conclusion; stalemate and the anticipation of it become interesting possibilities if the actions and information structure are in fact conducive to ties. But, with nonsymmetry as our philosophy, we do not need to be obsessed with the possibility of ties.

Again, if one player can make an offer and destroy communication, he may thereby win the ensuing tacit game by having provided the only extant offer that both players can converge on when they badly need to concert their choices later during the final tacit stage. To be sure, we can consider what happens when identical capacities for destruction of communication are present, and both players must recognize that they may simultaneously destroy communication without getting messages across; but this interesting case seems to be a special one, not the general case.

In summary, the perfectly "moveless" or "move-symmetrical" cooperative game is not a fruitful general case, but a limiting case that may degenerate into an ordinary tacit game. The cooperative game is rich and meaningful when "moves" are admitted; and much of the significance of the moves will vanish if complete symmetry in their availability to the players is stamped into the definition of the game. It is the moves that are interesting, not the

---

[12] It could be argued at this point that the expected value of the game is still symmetrically divided between the players, and that the analyst may consequently still view the game as symmetrical in terms of average outcomes. But if he does so he commits himself to a minimum of insight into the game and the way the game will be played.

game without moves; and it is the potential asymmetry of the moves that makes them most interesting.

Symmetry is not only commonly imposed on the move-structure of games but adduced as a plausible characteristic of the solution of the game or of the rational behavior with which the solution must be consistent. Nash's theory of the two-person cooperative game explicitly postulates symmetry, as does Harsanyi's. The symmetry postulate is certainly expedient; it often permits one to find a "solution" to a game and to stay — if he wishes to — within the realm of mathematics. There are few similarly potent concepts that compete with it as bases for solving a game. But the justification for the symmetry postulate has not been just that it leads to nice results; it has been justified on grounds that the contradiction of symmetry would tend to contradict the rationality of the two players. This is the underpinning that I want to attack.

What I am going to argue is that, though symmetry is consistent with the rationality of the players, it cannot be demonstrated that asymmetry is inconsistent with their rationality, while the inclusion of symmetry in the *definition* of rationality begs the question. I then want to offer what I think is *an* argument in favor of symmetrical solutions, an argument that tends to make symmetry but one of many potential influences on the outcome with no prima facie claim to pre-eminence.

Explicit statements of the relation between symmetry and rationality have been given by John Harsanyi. He says, "The bargaining problem has an obvious determinate solution in at least one special case: viz., in situations that are completely symmetric with respect to the two bargaining parties. In this case it is natural to assume that the two parties will tend to share the net gain equally since neither would be prepared to grant the other better terms than the latter would grant him." [13] In a later paper he refers to the symmetry axiom as the "fundamental postulate"

---

[13] Harsanyi, 147. He goes on to say, "For instance, everybody will expect that two duopolists with the same cost functions, size, market conditions, capital resources, personalities, etc., will reach an agreement giving equal profits to each of them."

and says, "Intuitively the assumption underlying this axiom is that a rational bargainer will not expect a rational opponent to grant him larger concessions than he would make himself under similar conditions." [14]

Now this intuitive formulation involves two postulates. First, that one bargainer will not concede more than he would expect to get if he himself were in the other's position. Second, that the only basis for his expectation of what he would concede if he were in the other's position is his perception of symmetry.

The intuitive formulation, or even a careful formulation in psychological terms, of what it is that a rational player "expects" in relation to another rational player, poses a problem in sheer scientific description. Both players, being rational, must recognize that the only kind of "rational" expectation they can have is a fully shared expectation of an *outcome*. It is probably not quite accurate — as a description of the psychological phenomenon — to say that one expects the second to concede something or to accept something; the second's readiness to concede or to accept is only an expression of what he expects the first to accept or to concede, which in turn is what he expects the first to expect the second to expect the first to expect, and so on. To avoid an "ad infinitum" in the descriptive process, we have to say that both sense a shared expectation of an *outcome*; one's "expectation" is a belief that both identify the *same* outcome as being indicated by the situation, hence as virtually inevitable. Both players, in effect, accept a common authority — the power of the game to dictate its own solution through their intellectual capacity to perceive it —

[14] The full quotation deserves to be given: "What the Zeuthen-Nash theory of bargaining essentially proposes to do is to specify what are the expectations that two rational bargainers can consistently entertain as to each other's bargaining strategies if they know each other's utility functions. The fundamental postulate of the theory is a symmetry axiom, which states that the functions defining the two parties' optimal strategies in terms of the data (or, equivalently, the functions defining the two parties' final payoffs) have the same mathematical form, except that, of course, the variables associated with the two parties have to be interchanged. Intuitively the assumption underlying this axiom is that a rational bargainer will not expect a rational opponent to grant him larger concessions than he would make himself under similar conditions." (Harsanyi, "Bargaining in Ignorance of the Opponent's Utility Function," Cowles Foundation Discussion Paper No. 46, December 11, 1957, quoted by permission of the author.)

and what they expect is that they both perceive the same solution.[15]

In these terms the first (explicit) part of the Harsanyi hypothesis might be rephrased: that there is, in any bargaining-game situation (with perfect information about utilities), a particular outcome such that a rational player on either side can recognize that any rational player on either side would recognize it as the indicated "solution." The second (implicit) part of the hypothesis is that the particular outcome so recognized is determined by mathematical symmetry. The first we might call the "rational-solution" postulate; it is the second that constitutes the "symmetry" postulate.

The question now is whether the symmetry postulate is *derived* from the players' rationality — the rationality of their expectations — or must rest on other grounds. If it rests on other grounds, what are they and how firm is the support?

To pursue the first question, whether symmetry can be de-

---

[15] Viewed in this way, the intellectual process of arriving at "rational expectations" in the full-communication bargaining game is virtually identical with the intellectual process of arriving at a coordinated choice in the tacit game. The actual solutions might be different because the game contexts might be different, with different suggestive details; but the nature of the two solutions seems virtually identical since both depend on an agreement that is reached by *tacit* consent. This is true because the explicit agreement that is reached in the full-communication game corresponds to *a priori* expectations that were reached (or in theory could have been reached) jointly but independently by the two players before the bargaining started. And it is like a tacit *agreement* in the sense that both can hold confident rational expectations only if both are aware that both accept the indicated solution in advance as *the* outcome that they both know they both expect.

There is a qualification to this point. With full information about each other's value systems and a homogeneous set of gains to be divided, there may be an infinity of equivalent solutions, all yielding the same values to the two players, but no difficulty in agreeing on an arbitrary choice among this indifferent set. But tacit bargaining often requires a further degree of coordination, namely, a coordinated choice even among equivalent divisions of the gains. Negotiation over a boundary line in homogeneous territory is thus different from the simultaneous dispatch of troops to take up positions representing claims (as in Question 6 on page 62); such claims may overlap and cause trouble even though the terrain values claimed are consistent. Thus the coordination problem is different; and there is no *a priori* assurance that the solution to the tacit game (or to games with somewhat incomplete communication, information, and so forth) would be in the set of equivalent solutions to the fully explicit game.

duced from the rationality of the players' expectations, we can consider the rationality of the two players jointly and inquire whether a jointly expected nonsymmetrical outcome contradicts the rationality postulate. If two players confidently believe they share, and do share, the expectation of a particular outcome, and that outcome is not symmetrical in a mathematical sense, can we demonstrate that their expectations are irrational, and that the rationality postulate is contradicted? Specifically, suppose that two players may have $100 to divide as soon as they agree explicitly on how to divide it; and they quite readily agree that A shall have $80 and B shall have $20; and we know that dollar amounts in this particular case are proportionate to utilities, and the players do too. Can we demonstrate that the players have been irrational?

We must be careful not to make symmetry part of the *definition* of rationality; to do so would destroy the empirical relevance of the theory and simply make symmetry an independent axiom. We must have a plausible definition of rationality that does not mention symmetry and show that asymmetry in the bargaining expectations would be inconsistent with that definition. For our present purpose we must suppose that two players have picked $80 and $20 by agreement and see whether we can identify any kind of intellectual error, misguided expectations, or disorderly self-interest, on the part of one or both of them, in their failure to pick a symmetrical point.

Specifically, where is the "error" in B's concession of $80 to A? He expected — he may tell us, and suppose that we have means to check his veracity (a modest supposition if full information of utilities is already assumed!) — that A would "demand" $80; he expected A to expect to get $80; he knew that A knew that he, B, expected to yield $80 and be content with $20; he knew that A knew that he knew this; and so on. A expected to get $80, knew that B was psychologically ready because he, B, knew that A confidently expected B to be ready, and so on. That is, they both knew — they tell us — and both knew that both knew, that the outcome would ineluctably be $80 for A and $20 for B. Both were correct in every expectation. The expectations of each were internally consistent and consistent with the other's.

We may be mystified about *how* they reached such expectations; but the feat claims admiration as much as contempt. The "rational-solution" postulate is beautifully borne out; the game seems to have dictated a particular outcome that both players confidently perceived. If, at this point, we feel that we ourselves wouldn't have perceived the same outcome, we can conclude that one of four hypotheses is false: (1) the rational-solution postulate, (2) the rationality of A and B, (3) our own rationality, (4) the identity (in all essential respects) of the game that we introspectively play with the game that A and B have just played. But we cannot, on the evidence, declare the second to be the false one — the rationality of A and B.

Note that if B had insisted on $50, or if A had been content to demand $50, claiming to be rational and arguing in terms of confidence in a shared expectation of that outcome, both players would have been in "error" and we could not tell, on the evidence, which one was irrational or whether they both were. Unless we made symmetry the definition of rationality we could only conclude that at least one of the players was irrational or that the rational-solution postulate did not hold. What we have is at best a single *necessary* condition for the irrationality of both players jointly; we have no sufficient condition, and no necessary condition that can be applied to a single player.

Nor can we trip them up if we ask them how they arrived at their expectations. Any grounds that are consistent would do, since any grounds that each expects the other confidently to adopt are grounds that he cannot rationally eschew. Consistent stories are all they need; and if they say that a sign on the blackboard said A-$80, B-$20, or that they saw in a bulletin that two other players, named A' and B', split $80-$20, and that they confidently perceived that this was clear indication to both of them of what to expect — that this was the only "expectable" outcome — we cannot catch them in error and prove them irrational. They may be irrational; but the evidence will not show it.

There is, however, a basis for denying my present argument. Since I have not actually applied an independent test of rationality to two players, given them the game to play, and observed the 80:20 split that I just mentioned, but have only posed it as a possibility to see whether it would imply irrationality *if* it occurred, one might object that it could not occur. And the argument would rest on the problem of coordination; it would run as follows.

If two players jointly expect *a priori* the same outcome, and confidently recognize it as their *common* expectation, they must have the intellectual power to pick a particular point in common. If the whole $100 can be divided to the nearest penny, there are 9,999 relevant divisions to consider, one of which would have to be picked simultaneously but separately by both players as their expectations of the outcome. But how can two people concert their selections of one item out of 9,999, in the sense that their expectations focus or converge on it, except with odds of 9,999 to 1 against them? The answer must be that they utilize some trick, or clue, or coordinating device that presents itself to them. They must, consciously or unconsciously, use a selection procedure that leads to unique results. There must be something about the point they pick that distinguishes it — if not in their conscious reasoning, at least in our conscious analysis — from the continuum of all possible alternatives.

Now, is it possible for two rational players, through anything other than sheer coincidence or magic, to focus their attention on the same particular outcome and each "rationally" be confident that the other is focussed on the same outcome with the same appreciation that it is mutually expected? And, if so, how can they?

The answer is that they can, as demonstrated in Chapter 3. They may use any means that is available: any clue, any suggestion, any rule of elimination that leads to an unambiguous choice or a high probability of concerted choice. And one of these rules, or clues, or suggestions, is mathematical symmetry.[16]

---

[16] The basic intellectual premise, or working hypothesis, for rational players in this game seems to be the premise that *some* rule must be used if success is to exceed coincidence, and that the best rule to be found, whatever its rationalization, is consequently a rational rule. This premise would support, for example, Nash's model that views an "unsmoothed" tacit game as the limit of a "smoothed" game as the smoothing approaches zero. While this view of the unsmoothed game is in no sense logically necessary, it is a powerfully suggestive one that can, in the absence of any better rationale for converging on a single point, command the attention of players in need of a common choice.

In a game that has absolutely no details but its mathematical structure, in which no inadvertent contextual matter can make itself appreciated by a player as something that the other can appreciate too, there may be nothing to work on but a continuum of numbers. And all the numbers can be sorted according to whether they correspond to symmetrical or asymmetrical divisions. If all numbers but one represent an asymmetrical split, then sheer mathematical symmetry is a sufficient rule and a supremely helpful one in concerting on a common choice. And it may be possible to set up a game in such sanitary fashion, suppressing the identity of players and all contextual details, that there is literally no other visible basis for concerting unless impurities creep in.[17]

In other words, mathematical symmetry may focus the expectations of two rational players because it does — granted the other assumed features of the game, like full information on each other's utility systems — provide one means of concerting expectations. Whether it is a potent means may depend on what alternatives are available.

That there are other means of concerting, including some that may substantially outweigh the notion of symmetry, seems amply demonstrated by the experiments in Chapter 3. So it is demonstrably possible to set up games in which mathematical sym-

---

The limiting process provides a clue for picking one of the infinitely many equilibrium points that actually exist in the unsmoothed game. Of course, the premise equally supports any other procedure that produces a candidate for election among the infinitely many potential choices.

[17] In this view, the theory of Nash (leading to the maximum-utility-product solution) is a response to the fact that even in the realm of mathematics there are offhand too many types of uniqueness or symmetry to provide an unambiguous rule for selection, hence a need to adduce plausible criteria (axioms) sufficient to yield an unambiguous selection. Braithwaite's theory can be characterized the same way. The fact that the two solutions conflict implies that mathematicians may not have a sufficiently common mathematical aesthetic to satisfy the first part of the Harsanyi postulate, that is, to coordinate their expectations on the same outcome. (R. B. Braithwaite, *Theory of Games as a Tool for the Moral Philosopher* [Cambridge, England, 1955]; Braithwaite's solution is described in Luce and Raiffa, *Games and Decisions*, 145ff.) Braithwaite's construction of the problem as a one-person arbitration problem, and Luce and Raiffa's reformulation of Nash's theory in terms of arbitration rather than strategy (pages 121–154), seem to emphasize that *intellectual coordination* is at the heart of the theory. A legalistic solution requires *some* rationalization of a unique outcome; pure casuistry is helpful if the alternative is vacuum.

metry does provide the focus for coordinated expectations, and demonstrably possible to set up games in which some other aspect of the game focusses expectations. (These other aspects are commonly not contained in the mathematical structure of the game but are part of the "topical content"; that is, they usually depend on the "labeling" of players and strategies, to use the term of Luce and Raiffa mentioned in Chapter 4.)

I have no basis for arguing with what force, or in what percentage of interesting games, mathematical symmetry does dominate "rational expectations." But I think that the status of the symmetry postulate is qualitatively changed by the admission that symmetry has competitors in the role of focussing expectations. For, if it were believed that rational players' expectations could be brought into consistency only by some mathematical property of the payoff function, then symmetry might seem to have undisputed claim, particularly if it is possible to find a unique definition of symmetry that meets certain attractive axioms. But if one has to admit that other things — things not necessarily part of the mathematical structure of the payoff function — can do what symmetry does, then there is no a priori reason to suppose that what symmetry does is 99 percent or 1 percent of the job. The appeal of symmetry is no longer mathematical, it is introspective; and further argument is limited to the personal appeal of particular focussing devices to the game theorist as game player, or else to empirical observation.

Thus a normative theory of games, a theory of strategy, depending on intellectual coordination, has a component that is inherently empirical; it depends on how people can coordinate their expectations. It depends therefore on skill and on context. The rational player must address himself to the empirical question of how, in the particular context of his own game, two rational players might achieve tacit coordination of choices, if he is to find in the game a basis for sharing an a priori expectation of the outcome with his partner. The identification of symmetry with rationality rests on the assumption that there are certain intellectual processes that rational players are incapable of, namely, concerting choices on the basis of anything other than mathematical symmetry, and that rational players should know this. It is an empirical question whether rational players can actually do what such a theory denies they can do and should consequently ignore the strategic principles produced by such a theory.[18]

An introspective game, which could be submitted to experiment, may illustrate the point. Imagine a game's potential payoffs as consisting of all the points on or within some boundary in the upper-right quadrant relative to a pair of rectangular coordinates.

[18] It is interesting that in demanding a symmetrical solution to an ostensibly symmetrical tacit game, Luce and Raiffa dismiss the two most promising candidates. They consider (*Games and Decisions*, 90–94) a matrix,

	I	II
i	1 / 2	−1 / −1
ii	−1 / −1	2 / 1

and note that it has pure-strategy equilibrium points in the upper-left and lower-right corners. These are ruled out on grounds that "whatever rationalization I give for either i or ii there is, by the symmetry of the situation, a similar rationalization for player 2, and so it seems inevitable that we both lose." (I have substituted i and ii for their designations.) They then look at a pair of maximin strategies, which are unsatisfactory because they do not produce an equilibrium point, and a minimax strategy which they find even inferior. But the important question is whether players who are both rational and imaginative are quite as impotent as Luce and Raiffa insist. Can players correlate strategies without communicating? This an empirical question; the experiments of Chapter 3 give an affirmative answer, or at least indicate that in particular cases the answer may be yes. Offhand it may seem hard for them to concert on a nonsymmetrical pair of strategies. But much the hardest part is just recognizing that they have to; the question of how to do it then becomes a practical matter. They must jointly and tacitly find a clue to the concerting of their choice. Of course, a nonsymmetrical solution in the above matrix is a discriminatory one; it quite arbitrarily condemns one of the players to a smaller gain than the other for reasons that may seem purely accidental or incidental. But we have to suppose that a rational player can discipline himself to accept the lesser share if the clue points that way. Only a discriminatory clue can point to a concerted choice; to deny the discrimination is to deny the premise that a clue can be jointly found and jointly acted on in the interest of an outcome that is jointly far superior to any symmetrical outcome. Luce and Raiffa conclude their discussion of this particular game with the remark that "although this seemingly innocuous game possesses some symmetries it is difficult to see how to exploit them." But the real key to this seemingly innocuous game is that it may, particularly when presented in a context, possess some *asymmetries*; and the object is to exploit them. See also pp. 298 ff.

Let us — whether or not we are strongly attracted to the symmetry postulate, and whether or not we are especially attracted to the particular symmetry of the Nash solution — put ourselves in a frame of mind congenial to accepting the "Nash point" as the rational outcome of an explicit bargaining game.[19] Consider now some variants of this game.

[19] The solution proposed by J. F. Nash for bargaining games in which both players have perfect knowledge of their own and each other's utility systems (subjective valuations) is the outcome that maximizes the *product* of the two players' utilities. If all possible outcomes are plotted on a graph whose rectangular coordinates measure the utilities that the two players derive from them, the solution is a unique point on the upper-right boundary of the region. (The point is unique because, if there were two, the two could be joined by a straight line representing available alternative outcomes achievable by mixing, with various odds, the probabilities of the original two outcomes; and points on the line connecting them would yield higher products of the two players' utilities. In other words, the region is presumed convex by reason of the possibility of probability mixtures, and a convex region has a single maximum-utility-product point, or "Nash point.")

A distinguishing feature of this particular "solution" is that it is independent of the exchange rate between the two players' utility scales; it is, in other words, invariant with respect to any fixed weights that we might attach to their respective utilities. And it meets some other conditions, notably including the condition that for any pair of fixed weights (or any exchange rate) relating the two players' utility scales that yields a *symmetrical* region, the upper-right midpoint is the solution; that is, the best point symmetrical as between the two players is the solution. (It is the only solution that does meet all of the specified conditions; Nash showed that any solution meeting his conditions must lead to the outcome that entails the maximum product for the two players' utilities.) For our present purpose we may take this symmetry requirement as the generic characteristic of the solution, and think of the other conditions (axioms) as serving to refine the crude notion of symmetry to the point where a unique solution is guaranteed. See the earlier references (p. 267) to Nash, Harsanyi, and Luce and Raiffa; see also the excellent elucidation of the Nash theory, with criticism, by Robert Bishop, "The Nash Solution of Bilateral Monopoly and Duopoly," to be published. And for an application of the "Nash point" to the theory of arbitration, see Layman E. Allen, "Games Bargaining: A Proposed Application of the Theory of Games to Collective Bargaining," *Yale Law Journal*, 65:660 (April, 1956).

Incidentally, it may deserve to be emphasized that the Nash theory is not just one that does not *need* a means for comparing two players' utility scales — one that, being independent of interpersonal utility comparisons, can get along without them. Rather, since it uses the arbitrariness of the utility exchange rate as a fundamental principle, the theory must be taken to *depend* on the inherent incommensurability of utilities. If the two players' utility scales could in principle be compared, though with difficulty, the Nash theory would not seem an attractive means of obviating difficult comparisons. If in principle utilities were commensurable, there would be little virtue in a theory

First, we are to play the same game in its tacit form. Each of us picks a value along his own axis, and if the resulting point is on or within the boundary, we get the amounts (utilities) denoted by the coordinates we pick. I conjecture that, in the frame of mind I have asked for — a frame of mind that made the Nash point appeal to us in the explicit-bargaining game — we should probably pick the Nash point. Without asking precisely why, let us go on to another variant of the game. This variant is tacit too; but it differs in that we get nothing unless the point whose coordinates we pick is *exactly* on the boundary. We get nothing unless we exhaust the available gains. Caution gets us nowhere; each must choose exactly as the other expects him to. I propose that in our present frame of mind we ought to take the Nash point.

Finally, consider another variant. We are shown the diagram of the game that has just been played and told that we are now to be perfect partners, winning and losing together. Conscious of the fact that our present game is modeled on a bargaining game we are to pick, without communicating, coordinates of a point that lies exactly on the boundary. If we do, we both win prizes — the same prizes no matter what point we succeed in picking together — and if we fail to pick a point on the boundary we get nothing. In this pure coordination game, I conjecture again that we should (would) in our present frame of mind pick the Nash point.

Why? Simply because we need some rationalization that leads to a unique point; and in the context, the bargaining analogy provides it. Unless there is a sharp corner (which is then likely to be the Nash point anyway); or a simple mid-point as when the

---

that relies, in reaching a solution, on the principle of incommensurability. And, while the present-day conceptual bases of game theory and of economic theory seem incompatible with interpersonal utility comparisons, the notion of *arbitration* may not be. Economic theory finds it convenient to use a notion of utility that makes utility theory correspond to choice theory, so that one can get "welfare economics" as a free by-product of a theory of economic choice. But if one were to forego this correspondence, for purposes of deriving principles of arbitration, one might be led either to an attempt to measure "utility" in some psychological or physiological way, or to establish legalistically some convention for making a comparison — a convention that, though arbitrary, were compatible with the social purpose of arbitration

boundary is a straight line or circular arc (which again coincides with the Nash point); or some especially suggestive form that seems to point towards a particular point; or unless there is an impurity (such as a dot on the boundary, from a printer's error, or a single point whose coordinates are whole numbers, and so forth), we may be led to search for a "unique" definition of symmetry to fall back on. And Nash-type symmetry is as plausible as any I can think of — not as simple as some (like the intersection with a 45° line from the origin of the diagram and others of that ilk), but less ambiguous on its own level of sophistication.

And, if the Nash point appeals to us powerfully in the bargaining game, it must do so because we are confident that it appeals equally to our partner who in turn we believe to be aware that our views coincide. It must therefore appeal to us in the pure-coordination game as a unique point that the partner will consider to be obviously obvious.

What does this prove or suggest? I am not arguing for the Nash point. I am arguing rather that the appeal of the Nash point to a game theorist (as introspective game player) may be the reverse of the sequence I have just run through. It may be the focal quality of the Nash-point in the pure coordination game — the unequivocal usefulness of a uniquely defined symmetry concept, when no nonmathematical impurities are available to help — that makes it a controling influence in the tacit and terribly cooperative boundary-line variant of the game; that in turn makes it a reliable guide in the less demanding tacit bounded-area variant of the game; and that in turn takes the heart out of any player in the explicit bargaining game who might hope that expectations could focus anywhere else.

In other words, by postulating the *need for coordination of expectations*, we seem to have a theoretical basis for something like the Nash axioms. What a theory like Nash's needs is the premise that a solution exists; it is the observable phenomenon of tacit coordination that provides empirical evidence that (sometimes) rational expectations can be tacitly focussed on a unique (and perhaps efficient) outcome, and that leads one to suppose that the same may be possible in a game that provides nothing but mathematical properties to work on. The Nash theory is vindication of this supposition — complete vindication if it dominates all competing mathematical solutions in terms of mathematical esthetics. The resulting focal point is limited to the universe of mathematics, however, which should not be equated with the universe of game theory.

RE-INTERPRETATION OF A SOLUTION CONCEPT FOR "NONCOOPERATIVE" GAMES

The pure common-interest game, or coordination game, may add insight into the reasoning behind certain solution concepts in game theory, particularly that of *solution in the strict sense* for the "noncooperative" game. By "reasoning that lies behind these concepts" I mean the reasoning that is imputed to the rational players to whom the concepts should appeal.[1]

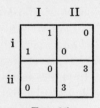

FIG. 25

The tacit games represented in Figs. 25 and 26 are said to have a *solution in the strict sense*. (In Fig. 26 a choice of either second or third strategy for each player constitutes the solution.) The definition of such a solution, given by Luce and Raiffa, is as follows: "A non-cooperative game is said to have a *solution in the strict sense* if: (1) There exists an equilibrium pair among the jointly admissible strategy pairs. (2) All jointly admissible equilibrium pairs are both interchangeable and equivalent."[2]

---

[1] "Noncooperative" is the traditional name for the game without overt communication. Unfortunately it may suggest that cooperation is absent when communication is absent. As indicated in Chapters 3 and 4, cooperation — reciprocated and taken for granted by each side — is an essential element, even a dominant element, in many tacit nonzero-sum games.

[2] *Games and Decisions*, p. 107f. This particular solution concept is akin to,

	I	II	III
i	1 / 1	0 / 0	0 / 0
ii	0 / 0	3 / 3	3 / 3
iii	0 / 0	3 / 3	3 / 3

Fig. 26

An *equilibrium pair* is a pair of strategies for the two players such that each is the player's best strategy (or as good as any other) that can be coupled with the other's. A *jointly admissible* strategy pair is a pair that is not jointly dominated by another pair; that is, it yields a pair of payoffs that are not both inferior to the payoffs in some other cell. Equilibrium pairs are *equivalent* if, for each player separately, they yield equal payoffs; equilibrium pairs are *interchangeable* if all pairs formed from the corresponding strategies are also equilibrium points. (They are therefore equivalent and interchangeable only if all pairs formed from the corresponding strategies are equivalent.) Thus the strategy pairs (ii, II), (iii, III), (ii, III), and (iii, II) in Fig. 26 denote equivalent, interchangeable, jointly admissible equilibrium pairs.

Luce and Raiffa, immediately after this definition, add the following comment, which can serve as our point of departure: "The second condition prohibits *confusion* in the case of non-unique jointly admissible equilibrium pairs." (My italics.)

It is precisely this problem of *confusion*, or *ambiguousness*, that was at the heart of the coordination game in Chapter 3. The game in Fig. 27 does not have a *solution in the strict sense*. The second and third strategies for the two players are not interchangeable and equivalent — they do not yield equivalent pairs in all four combinations. There is no difference of interest between the two players in their choice of strategies; there is simply cause for confusion. In Fig. 1 they know exactly what

but distinct from, that proposed by J. F. Nash in 1951. For a comparison of several related solution concepts see Chap. 5 of Luce and Raiffa, and J. F. Nash, "Non-cooperative Games," *Annals of Mathematics*, 54:286–295 (1951).

	I	II	III
i	1 1	0 0	0 0
ii	0 0	3 3	0 0
iii	0 0	0 0	3 3

FIG. 27

strategies to choose; in Fig. 2 they know as well as they need to; in Fig. 3 they do not. Failure to coordinate in Fig. 3 condemns them to zero apiece, and without a clue to coordination they may be supposed to have a fifty-fifty chance of winning 3 apiece, for an expected value of 1.5.

Why is it that (ii, II) is the indicated solution in Fig. 25, rather than (i, I)? An offhand answer is that the payoff is better for (ii, II) than for (i, I). But this is only part of the answer. Another part emerges if we look at Fig. 28, which is like Fig. 25

	I	II
i	9 9	0 0
ii	0 0	10 10

FIG. 28

in preference ordering but different in absolute strengths of preference. In Fig. 28 it looks as though the important thing is not to achieve 10 rather than 9, but 9 or 10 rather than zero. Roughly speaking, the two equilibrium pairs are nearly equivalent but not interchangeable; and though the players may be little concerned about whether they get 9 or 10 they are very much concerned not to get zero. Their main interest is to avoid "confusion."

They need to find some clue, or rule, or instruction to coordinate their choices. In a game as abstract as the matrix in Fig.

28, there is little to guide them but the numbers; and between the alternative rules of picking the lesser pair or the greater, the latter probably has more plausibility. We might ask how much it is worth to the players to have an extra dollar attached to (ii, II) by comparison with (i, I); it is worth a great deal as a signaling device and just a little as extra money. It is the difference between 9 and 10 that makes it possible to coordinate choices. In Fig. 29, if we suppose that they can find no rule

FIG. 29

for coordination, their expected value is presumably 5 apiece.

(Actually the game in Fig. 29 *if presented in the matrix as shown* may not cause difficulty. The empirical results of Chapter 3 imply that it need not. A specific matrix permits left-right, upper-lower, first-last-middle distinctions. For our present purpose, we must suppose that the strategies occur to the players in such form and with such labels that rational players are intellectually incapable of ordering them unambiguously. A completely foolproof or geniusproof clueless game would presumably have to have scrambled labels and a perfectly symmetrical set of payoffs. Incidentally, a tacit game with infinitely many strategies apparently has no "pure" form; an infinity of strategies could only be presented to the players by means of a generating formula, and any generating formula is likely to offer the players some means of ordering the strategies.)

The situation may not be very different if we suppose that the strategy pair (ii, II) is underlined, printed bold face, has arrows pointing toward it, or has a footnote saying that in case of confusion the management suggests a choice of (ii, II). What the players need is *some* signal to coordinate strategies; if they cannot find it in the mathematical configuration of the payoffs, they can look for it anywhere else. And strategies may occur in such fashion, or with such labels or connotations, as to provide a potential basis for ordering them or sorting them that rational players find useful.[3]

The suggestion of this appendix, then, is that an important property enjoyed by a "solution in the strict sense" — a reason why rational players might select it — is a signaling power, a means of tacit communication, that is available to the two players to facilitate their tacit cooperation when failure to coordinate choices would be serious. This is of course not the only significant property of such a solution; but it may be an important part of the rationale for a player's choosing it.

Another way to make this point is that we could, in games like those presented in this paper, prescribe communication arrangements with certain communication costs and analyse the games to see whether communication is worth the cost and what messages sent over what channels would constitute the "solution." The "clues" under discussion in this paper would then appear to be so much free communication to be taken advantage of; and it is an empirical question what free communication a rational player should be able to find and take for granted. Just as esthetic or syntactic constraints on a language help to eliminate garbles in a badly transmitted message, esthetic or dramaturgical constraints, casuistic or geometric constraints, can help to eliminate ambiguousness in a situation where tacit concerted choice is required.

The point can be pressed further. Consider the game in Fig. 30. Again assume that the strategies occur in a way that makes ordering them intellectually impossible for rational players, specifically, not in the form of a particular square matrix, not

[3] The type of "rationality" or intellectual skill required in these games is something like that required in solving riddles. A riddle is a context in which one is invited to search for a clue, the rules being that the clue must not be too hard to find nor too easy. (One must at least be able to recognize that he should have got it, when it is pointed out to him.) A riddle is essentially a two-person problem; the methodology of solution depends on the fact that another person has planted a message that in his judgment is hard to find but not too hard. In principle one can neither make up nor solve riddles without empirical experience; one cannot deduce *a priori* whether a rational partner can take a hint. "Hint theory" is an inherently empirical part of game theory.

	I	II	III	IV
i	10 / 10	0	0	0
ii	0	10 / 10	0	0
iii	0	0	9 / 9	0
iv	0	0	0	10 / 10

Fig. 30

labeled with numbers or letters, or — if they are labeled — with the labels scrambled separately for the two players. There it would appear that if no better means of coordinating can be discerned, the "solution" may be the strategy pair (iii, III) with payoffs of 9 apiece. This is the least desired among the equilibrium points, but it enjoys uniqueness while the others offer confusion; it provides a clue to concert choices. In terms of the *payoff* structure alone (that is, without introducing "labels," prefabricated matrices, or any other details outside the pure quantitative structure of the game), it is hard to see that this solution is much less, if at all less, compelling than the one in Fig. 31,

	I	II	III	IV
i	9 / 9	0	0	0
ii	0	9 / 9	0	0
iii	0	0	10 / 10	0
iv	0	0	0	9 / 9

Fig. 31

although the latter meets the Luce-Raiffa definition and the former contradicts it.[4]

	I	II
i	10, 10	0, 0
ii	0, 0	10, 10

FIG. 32

The games in Figs. 32 and 33, neither of which has a solution in the strict sense, seem to represent the same point. It "looks as though" the players have an argument for choosing (ii, II) in Fig. 33. One argument might be that, in the absence of any way of knowing whether to aim for (i, I) or (ii, II), one should consider what insurance he can fall back on. The row chooser gets nothing if he wrongly chooses the upper row, he gets 5 if he wrongly chooses the lower row, "wrong" meaning that he fails to rendezvous with his partner for 10. He might then choose the lower row arguing that he does so because he will at least get 5 if he does not get 10, and his chances of getting 10 are no worse with this choice. Perhaps this is all that "rationality" requires of him; but it might be more perceptive to reason as follows.

	I	II
i	10, 10	0, 5
ii	5, 0	10, 10

FIG. 33

"Comparing just (i, I) and (ii, II) my partner and I have no way of concerting our choices. There must be some way, however, so let's look for it. The only other place to look is in the cells (ii, I) and (i, II). Do they give us the hint we need

[4] Empirical evidence for these and similar games can readily be obtained for himself by any reader who wants to pursue the point.

to concert on 10 apiece? Yes, they do; they seem to "point toward" (ii, II). They provide either a reason or an excuse for believing or pretending that (ii, II) is better than (i, I); since we need an excuse, if not a reason, for pretending, if not believing, that one of the equilibrium pairs is better, or more distinguished, or more prominent, or more eligible, than the other, and since I find no competing rule or instruction to follow or clue to pursue, we may as well agree to use this rule to reach a meeting of minds."

In this case the players are not choosing their second strategies because 5 is preferable to 0. They have no serious expectation of getting 5. They are *using* the configuration of fives and zeros as a *clue* to coordinating actions. It is *useful* to the players — and each recognizes that the other recognizes that it is useful — to take note of where the fives are, but only as a step in the process of coordinating intentions. The tendency for the matrix in Fig. 33 to "converge" on (ii, II) is in principle the same as if the printed matrix had arrows pointing toward the lower-right corner, arrows with no logical role or authority other than the power of suggestion and hence the ability to coordinate expectations.[5]

CONFLICTING INTEREST

We can consider now the case of coordination mixed with conflict. Figures 34 and 35 portray games that have equilibrium points, two of them both jointly admissible, without a "solution in the strict sense" because the equilibrium pairs are neither equivalent nor interchangeable.

The coordination problem in the first of the two is apparently "insoluble" in its purely abstract form, that is, without labels on the strategies; there appears to be at best a random chance

[5] Assuming that a player does choose ii or II, it may be worthwhile to find an operational way of discriminating between motives for choosing it, even if only to make sure that the concept is operational. As between the two motives mentioned — the "insurance" motive and the "coordination-clue" motive — we might distinguish as follows. We offer a player alternative games like Fig. 33 that differ only in substituting values ranging from 0 to 9 for the 5's in that matrix, leaving the 10's and zeros as they are. We then ask him to "value" the games for us — to indicate how much he would pay for the opportunity to play the game with a live partner and real money payoffs. (Alternatively

I   II

	I	II
i	4 / 6	0 / 0
ii	0 / 0	6 / 4

FIG. 34

of achieving either of the jointly admissible (efficient) outcomes.[6] The second may not be insoluble. Each player would rather accept his "second-best" equilibrium point than fail to coordinate at all; they have a common interest in cooperating to find a clue to common choice. Why not take the clue contained in the other cells, which seems to point toward (ii, II)?[7]

we ask him how much he'd pay for the privilege of playing the different variants in place of the one with 5.) If his response is fairly insensitive to variations in that particular payoff as long as it is positive, and if nevertheless he attaches a high value to the game with some positive payoff and attaches something like a random-strategy expected value for the game with zeros as in Fig. 32, we can conclude that the lower-left and upper-right payoffs are mainly of interest to him as signals. If, for example, he bids $9.50 for a chance to play the game in Fig. 33 (implying, perhaps, a 90 percent expectation that Column will choose II), $8.65 for the game with 5 replaced by 1 (implying an 85 percent expectation of II), and $9.95 for the game with 5 replaced by 9 (implying a 95 percent expectation of II), and, finally, $5 for the game as in Fig. 32 (implying a random expectation as between I and II), we could conclude that the function, or value to the player, of the upper-right and lower-left payoffs is largely that of coordinating clue. If instead he bids amounts that imply probabilities between I and II that are invariant, or nearly so, with respect to the upper-right and lower-left payoffs, and particularly if he bids the arithmetic mean, the insurance interpretation would be indicated. (Note that the adjectives "upper-right" and "lower-left" are only author's shorthand here; they have no meaning to the player since we are considering the case of *unlabeled* strategies, which must not be presented in a square matrix, or with labels like "i" and "ii" — or, if they are, must have been labeled by a random process separate from the random process that allocated labels or positions for the other player. Specifically, Row must not know whether Column's matrix looks like Fig. 33 or instead has the columns interchanged with the low-value payoffs in upper-left and lower-right.)

[6] See the footnote on p. 286 for a discussion of a similar matrix when the premise of pure abstraction is relaxed.

[7] The game in Fig. 35 does have another equilibrium point, consisting of an 80:20 mixed strategy for row and a 40:60 mixture for column. It yields them payoffs of 3.6 apiece, and is therefore jointly dominated by the upper-left and lower-right cells.

	I	II
i	4 \ 6	3 \ 2
ii	2 \ 3	6 \ 4

FIG. 35

For one of the players this is not the most advantageous outcome, but beggars cannot be choosers when fortune gives the signals. What other clue is there? It might be equally *fair* to use the negative of this clue; just as it would be equally fair, if arrows pointed toward (ii, II) and away from (i, I), to treat the feathers as the signal rather than the arrowheads. But fairness cannot help; in fact it makes coordination impossible. If all clues are equally plausible in reverse, we are back to confusion. Only a *discriminatory* clue can point to a concerted choice, denying the discrimination is denying the premise that a clue can be found and acted on jointly to achieve an efficient outcome in the face of conflicting preferences.[8]

Here again the most potent clues may be those that we admit when we go beyond the mathematics of the payoff matrix. If we are driving toward the same intersection on perpendicular roads on a desert where no legal system determines right-of-way, and dislike and distrust each other and recognize that there is no moral obligation between us, the one approaching on the other's left may nevertheless still slow down to let the other through first, to avoid emergency stops at the intersection; and the other driver may anticipate this.[9] The conventional priority system lacks legal or moral force; but it is so expedient when coordination is needed that the one discriminated against may yield to its

[8] The power of similar mutually perceived signals seems to lie behind the concept of "psychological dominance" used by Luce and Raiffa to discuss the appeal in certain games of a jointly inadmissible equilibrium point. See *Games and Decisions*, pp. 109–10. See also the footnote on p. 286 for a comment on a similar game.

[9] A conflict-of-interest problem of this type — two cars approaching an asymmetrical narrow place in the road from opposite directions — was included in the questionnaire described in Chapter 3. The results bore out the general principle, but were omitted for brevity from Chapter 3.

discipline, recognizing that he should be grateful for an arbiter, even though it discriminates against him, and recognizing also that he is trapped by the other's acceptance of the signal and expectation that both will comply. By this reasoning, as developed in Chapter 3, the game in Fig. 34 may be soluble when presented in a *particular* matrix form to both players (that is, presented just as shown in Fig. 34), or when the winning strategy pairs are *labeled* "heads" and "tails," i, ii, I, and II, and so forth.

MANIPULATION BY A THIRD PARTY

Incidentally, all of these games requiring coordination, both those with conflicting preferences and those with preferences that coincide, might be substantially subject to the control or influence of a mediator. If we give a third player power to send messages to the original two tacit players, he is in a good position to help them; he is even in a good position to help himself if he gets a payoff that depends on the pair of strategies that the original two players choose. A benevolent mediator makes the pure common-interest game trivially easy; a mediator has an arbitrary power of justice in a game like that of Fig. 34; [10] a mediator is in a strong "third player" position in the game in Fig. 36, where the entry in parentheses is the payoff to the

	I	II	III	IV
i	6 (2) 5	0	0	0
ii	0	9 (3) 7	0	0
iii	0	0	5 (4) 6	0
iv	0	0	0	7 (1) 9

FIG. 36

[10] Recall problem no. 8 on p. 62 of Chapter 3, involving lost and found money and a self-appointed mediator.

mediator (or communication monopolist) who is in a position to give instructions — suggestive only, not authoritative — to the other two players.

INTERPRETATION OF THE PAYOFFS

As a final point it may be noted that, for the line of reasoning developed here, it does not matter whether we interpret the payoffs as objectively measurable entities, such as money or homogeneous goods, or as "utilities" in the sense now familiar in game theory. It does not depend on each person's knowledge of the strengths of the other's preferences, as long as the nominal payoffs are known. (If both the objective values and the utility values were known, and were not proportionate to each other, the "signals" might lose some force; the problem of confusion or ambiguousness would be aggravated.)

NUMBER OF PLAYERS

The discussion here has considered only two-person games, except for brief consideration of a third player who may be in a nontacit role. But the problem can be extended to any number of players, with the rewards depending either on unanimous choice or on some kind of majority or plurality choice or successful coalitions (somewhat analogous to the lines of the actual questionnaire procedure described in Chapter 3). The problem of ambiguousness may then become more serious, and the coordination aspect of the game may become even more relevant to the rationale of a "solution." It is probably in the realm of more-than-two-person games that coordination theory is most relevant of all, games involving the formulation of coalitions. Study of the signals and communication channels in coalition formation appears to be a fruitful meeting ground for game theory and sociology.

CONCLUSION

In summary, coordination-game theory suggests that the "solution in the strict sense" of a tacit nonzero-sum game is to be understood partly, and in some cases largely, by reference to its signaling qualities. Since other sources of signals may be present *even in the purely mathematical formulation of the game,* the particular qualities of the "solution in the strict sense" are but one of many potential determinants of a "rational solution." It is partly an empirical question, not solely a matter of deduction a priori, what signals can be appreciated.

INDEX